Marriage Fictions in Old French Secular Narratives, 1170–1250

T0347371

Medieval History and Culture

Volume 6

STUDIES IN
MEDIEVAL HISTORY AND CULTURE

edited by
Francis G. Gentry
Professor of German
Pennsylvania State University

A ROUTLEDGE SERIES

OTHER BOOKS IN THIS SERIES

MARRIAGE FICTIONS IN OLD FRENCH SECULAR NARRATIVES, 1170–1250

A Critical Re-evaluation of the Courtly Love Debate

Keith Nickolaus

Routledge
Taylor & Francis Group

LONDON AND NEW YORK

First published 2002 by
Routledge

Published 2013 by
Routledge
2 Park Square, Milton Park, Abingdon, Oxfordshire OX14 4RN
711 Third Avenue, New York, NY 10017

First issued in paperback 2014

Routledge is an imprint of the Taylor and Francis Group, an informa business

Library of Congress Cataloging-in-Publication Data

Nickolaus, Keith.
    Marriage fictions in Old French secular narratives, 1170–1250 : a critical
    re-evaluation of the courtly love debate / by Keith Nickolaus
        p. cm. — (Medieval history and culture ; vol. 6)
    Includes bibliographical references and index.
    1. French literature—To 1500—History and criticism. 2. Romances—History
    and criticism. 3. Courtly love in literature. 4. Marriage in literature. I. Title.
    II. Series

PQ221 .N53 2001
840.9'3543'09021—dc21

2001048164

ISBN 978-0-415-93722-1 (hbk)
ISBN 978-1-138-86860-1 (pbk)

Series Editor Foreword

Far from providing just a musty whiff of yesteryear, research in Medieval Studies enters the new century as fresh and vigorous as never before. Scholars representing all disciplines and generations are consistently producing works of research of the highest caliber, utilizing new approaches and methodologies. Volumes in the Medieval History and Culture series will include studies on individual works and authors of Latin and vernacular literatures, historical personalities and events, theological and philosophical issues, and new critical approaches to medieval literature and culture.

Momentous changes have occurred in Medieval Studies in the past thirty years in teaching as well as in scholarship. Thus the goal of the Medieval History and Culture series is to enhance research in the field by providing an outlet for monographs by scholars in the early stages of their careers on all topics related to the broad scope of Medieval Studies, while at the same time pointing to and highlighting new directions that will shape and define scholarly discourse in the future.

Francis G. Gentry

This work is dedicated to my family and my wife, all of whom helped and encouraged me along the way.

Table of Contents

Preface

Amidst a long-awaited and growing prosperity that gives rise to a lavish aristocratic culture centered around courts that are less bastions of military defense than they are centers of vast dynastic and political ambition as well as magnets for both learning and luxury and at a time when religious institutions are in a period of transition fueled both by reformist zeal, the inroads of secular wealth and power, and the simultaneous call for and disillusionment with the Crusades, a significant social and intellectual *renaissance* takes place that shakes the foundations of "tradition" and "doctrine" that centuries of patristic discipline and elitism had so thoroughly fortified. In the realm of intellectual and literary developments, however, the torrent of change remained tempered by a deep and conscious reverence for tradition. In what would become an oft-quoted depiction of the paradoxical temper of these new thinkers, who, by the way (and despite the weight of patristic prejudice) were in many cases women, one monastic writer proclaims, "We are but dwarfs on the shoulders of giants." These giants were not limited to St. Augustine and the other important patristic writers who laid the cornerstones of medieval religious thought; among them were also important and newly recovered pagan writers whose eloquent and intricate Latin verses and epic compositions did indeed make them seem as towering giants to the new clerical "philosophers" and "poets" of the twelfth century who were, after centuries of strict religious conformity, now venturing to think of themselves as individual and innovative thinkers and "novelists." Virgil demonstrated that a profane epic was not unable to rise beyond the level of frivolous distraction. Indeed, no later period of critical theory has yet invested the narrative, the "book" (be it sacred or secular), with more potential for meaning than did the tireless exegetes who toiled throughout the Middle Ages. While pagan medieval culture had maintained its own tradition of epic story telling to commemorate the deeds of great leaders and warriors, the idea that profane narratives could actually be crafted by poet-philosophers as works of literary and allegorical illumination and edification gave rise to a brave new world of literary creation and invention. Likewise, there was a new aristocratic and "courtly" audience—increasingly more educated and more refined than the rural warlords celebrated in traditional epics—who were ready for new forms of entertainment, expression, and indoctrination. Perhaps one of the most materially insignificant and yet most original manifestations to emerge from this strange crucible of tradition and innovation is the medieval *romance*. On one level, these narratives were mere flights of fancy written in *romans*—the term for the common Latin vernacular of the age, the vernacular of secular society (as opposed to the formal Church Latin of trained clergy). These were novelistic narratives, most often in simple verse form, and developed as rather liberal "translations" of Latin epics (the stark Trojan War chronicles of Dares and Dictys which Benoit de St.-Maure turns into a multi-volume, love and tragedy filled account of the Trojan War called the *Roman de Troie* which complements two other contemporary "translations" of classical material, the *Roman de Thebes* and the *Roman d'Eneas*.) Later "romances" would be drawn from more obscure material and were likely to be the product of more original authorial invention: The sentimental *Lais* of Marie de France; a number of tales of love and knighthood by Chretien de Troie; a modified and expanded retelling of the medieval epic *Girart de Roussillon*; various versions of the amorous tragedy *Tristan et Iseut*; Jean de Meun's *Guillaume de Dole ou le roman de la rose*; the anonymous and voluminous tale of the amorous knight *Partonopeus de Blois*—to only mention works and authors that will be mentioned or discussed in this study. Fanciful, on the one hand, these "romances" did, nevertheless, signal the rise of an entirely new literary culture and a new freedom of authorial individuality and expression that had little precedent in the previous centuries of patristic-oriented letters and learning. At the same time, these authors were not pretending to a radical "originality" or creating literary narratives *ex nihilo*. They were drawing on a wide array of past and present influences; on a variety of genres, traditions and languages. Hence there is a peril for the literary critic, for the contemporary reader. The untutored reader is ill prepared to see the boundaries and resonances of diverse traditions and rich exegetic layerings; the trained critic, on the other hand, is likely to look for too much on a single pathway of tradition, to equate a reference to a known tradition to the substance of that tradition or, as a serious modern "theorist," to give too much weight to his / her own modernity and to ignore or abuse the realities of medieval tradition.

For all of the above reasons, the new *romance* genre of the twelfth century lends itself properly to a broad and comprehensive comparative approach. I originally defined this topic of dissertation inquiry in an effort to resolve what had become a confusing, perplexing and often fruitless "debate" over the origins and the meaning of "courtly love" and courtly love motifs in medieval literature. Some critics wished to dispel the confusion by eliminating use of the term altogether. However, the new secular "romance" which I sought to study, no doubt showed an evident and substantial preoccupation with amorous themes for which there was no obvious precedence and on account of which one had to wonder how such an interest in profane themes was to be reconciled with the underlying Christian orientation of clerically trained authors and the Christian virtues they extolled. These authors clearly had an interest in sentimentality and courtship when they crafted these stories. Why? Furthermore, why would adulterous plots and subplots also have such a prominent role in connection with the edifying heroes and heroines of these tales? I became convinced that to approach the question, one would have to read these tales neither as Christian tales nor as heresy, neither from the angle of patristic thought nor from the angle of anti-Christian sentiment. Indeed, these were tales written by poets well versed in Christian texts and ideals, but who were writing for sophisticated *secular* patrons and for a secular, aristocratic public. These were writers and readers who were becoming increasingly interested in, aware of, and appreciative of non-biblical and non-patristic traditions—who now saw exegetic value in profane Latin and Celtic traditions as well as in the emerging lyrical and narrative writing in their own vernacular. Hence, any critical study would have to look frequently to Virgil and Ovid as well as to Celtic influences. Furthermore, one would have to look not only at intertextual parallels but to the larger question of reception, *i.e.,* how did twelfth-century writers view and understand the literary value and meaning of these pagan traditions? This last question seemed to me to be of key importance in accurately interpreting the proper value of profane motifs in this cultural milieu. Finally, could one find currents of social, intellectual, and religious thought and doctrine that would logically account for the specific modalities of love, adultery, and marriage so keenly portrayed in the emerging *romans*?

Such are the questions I try to answer and such the broader cultural developments and literary spectrums that I try to integrate into this study of "courtly love" as portrayed in secular fictions about love, marriage, and adultery. I also review and critique at the outset, for my own and for the reader's benefit, the tangled web of "courtly love" criticism that confronts all students of medieval literature and which I think was leading to a significant critical impasse. My belief is that this study sheds a helpful light on many perplexing questions posed by fictional depictions of love, adultery, marriage, and courtship and provides an important remedy by giving these questions the broad scope they deserve and require. By examining a broad chronological sample of the genre and by omitting consideration of neither close textual reading nor broad literary and intellectual history, my approach identifies clear and consistent narrative and cultural frameworks that give us reasonable and convincing insights into the *meaning* of "court-ly love" in the twelfth and thirteenth-century Old French *roman.*

This book represents my dissertation in its original form. Aside from this *Preface* and the substitution of an expanded and updated *Bibliography* for the original *Works Cited* section as well as the addition of the *Index,* no changes have been made.

I remain forever grateful to my professors and my family, without whom this undertaking would never have been possible.

K. N.
Berkeley
March 2001

Introduction

Marriage and courtship dramas have a preponderant role in Old French narrative. Indeed, the very emergence of a literature in the vernacular language (*roman*) coincides with the emergence of a genre (the *roman*) dominated by what we now call "romance" themes. To some extent, the representation of love, adultery, courtship, and marriage in the Old French *roman* reflects rhetorical and sentimental features of Provençal love poetry and its ethic of *fin'amor*. Many critics have attempted to view love and courtship themes in secular narratives as an extension of lyric *tropes*. And, as long as critics would continue to define *fin'amor* as an adulterous form of courtship, the seminal importance of the Old French *Tristan* would appear to confirm the close relation between lyric and narrative models. Often, however, the narrative courtships culminate in an amorous marriage. As a result, many marriage fictions have been perceived as expressing a reactionary attitude toward adultery, and the study of narrative themes has subsequently been implicated in and prejudiced by the perplexing debate over the meaning of "courtly love." This means, despite all that has been written about the troubadours and "courtly love," about the influence of Ovid and the "rules" of love set forth by Andreas Capellanus, that medievalist scholarship suffers nonetheless from a serious lack of consensus with regard to the larger significance of, and motivations behind, Old French courtly narratives.

In relation to the study of Old French courtly narratives specifically, the problem is twofold. First, the appreciation of these works has been limited or distorted by exaggerated views of the influence exerted by lyric genres and by the controversial and largely irrelevant discussions relating to the nature and origins of *fin'amor* expressed in the Provençal *canso*. Second, critics have failed to account in fundamental terms for the convergence of Christian and profane themes; they have failed to reconcile the profile of the twelfth-century author and *clerc*—a man or woman of letters, but a fundamentally religious man or woman living in a fundamentally religious age—with salient and sympathetic representations of adultery in secular narratives. Thus, while the Robertsonian school of thought, which saw all representations of secular love and courtship in terms of patristic attitudes and satirical irony, has rightfully been criticized for its monolithic views of medieval Christian culture and its own selective deceits, critics have not very well resolved Robertson's central problem with regard to the assumed attitudes of medieval writers and the juxtaposition of the sacred and profane in the twelfth-century *roman*. The study of Romanesque sculpture and iconography led Loomis to express the same central query. In reference to the Arthurian friezes sculpted on the cathedral at Modena in Italy (probably in the first quarter or first half of the twelfth century), Loomis says: "No such excuse could possibly be invoked to justify the intrusion of the theme of the rescue of Arthur's queen into a sacred edifice. If excuse there was, we do not know and in all probability never will know what it was" (35). In the end Loomis invokes the tired moral dichotomies that frequently underlie modern critical approaches to the Middle Ages. He speculates that the Modena sculpture reveals a departure from strict monastic morality (35–36). Clearly the secular motif does not seem to reflect any clerical irony, and the sanctity of the edifice can hardly be questioned, even less so than the orthodoxy of an anonymous twelfth-century author. One way around the Robertsonian dichotomy between *caritas* and *cupiditas* is to view profane motifs not as ironic (or cautionary) *exempla* but as instances of mere secular diversion, the attitude adopted by Loomis. As Loomis himself implies, however, the Modena sculpture does not readily lend itself to such a solution. If its value were that of popular entertainment, how would it find itself permanently inscribed upon the facade of such a hallowed sanctuary?

An alternative approach would consist in viewing the secular motifs as authentic religious *exempla*—*exempla* appealing to secular patrons but communicating a collective ethos as well, an ethos relating to a secular institution with little biblical and patristic memory, an institution such as marriage. Loomis himself points the way when he reflects not on iconography but on the testimony of twelfth century secular narrative, on the continuous process of synthesis and rationalization inherent to the development of Arthurian lore in Old French.[1] He comes close, in fact, to proposing what I am contending here, that literary phenomena might suggest an answer to his question, that such sculptures might reflect the rational syntheses of authors who adapted mythological and religious motifs to secular narratives as part of what is obviously a larger syncretic tradition. In fact, it is interesting to note that the figures on the Modena archivolt reflect not just any profane lore, but a specific relation to marital conflict, an interest in the abduction and sovereignty-bride motifs descending (in this case) from early Irish saga:

> It is only when we recognize that the sculpture represents a composite tale
> centering around the abduction of Arthur's queen, sometimes called
> Guenloie or Guendoloena, here called Winlogee, that the whole scene and
> practically every actor in it can be understood and accounted for. (Loomis
> 34)

Perhaps marriage fictions dominate Old French narrative because questions of marriage, courtship, and adultery had gained a profound resonance for reasons independent of the vogue in lyric love poetry. A fresh and
thorough study of marriage fictions based on a wide sampling of works
promises to yield valuable insights into both the "problem" of "courtly
love" and the "moral" questions raised by representations of adulterous
courtship.[2] Had medieval critics heretofore focused more critical attention
on the *relation* between epic and amorous elements in composite works
such as the *romans d'antiquité*, they would no doubt have seen that the
integral relation between marital and sovereignty rivalries witnessed in
Indo-European myths and sagas linked marriage and courtship themes to
questions of national and universal history. For the medieval author, Ovid
clearly demonstrated what was already inherent in the epic battles for the
Greek Helen and for Vigil's Lavinia—that love, alongside war, was a twin
pillar, or perhaps the primary mover itself, in the destiny of secular
individuals, the triumph of sovereign nations, and the progress of world
history.

But what about the *Christian* Middle Ages and the *Christian* profile of
the twelfth-century *clerc* and author? What motivations would lead the
intellectual elite, trained in cathedral or monastic schools, to explore secular history, Classical tragedy, and pagan mythology? We know that pagan
deities lived on through the revival of Classical learning in the twelfth century,[3] but what about the words of philosophers and poets? What necessary, edifying, or exemplary value might these have for the creation of
imaginative literature in a thoroughly Christian culture? As for orthodox
patristic doctrine and attitudes, the twelfth century coincides, in fact, with
a profound reform of traditional moral doctrine. For along with monastic
reforms in the eleventh and twelfth century came a more positive concept
of the marital bond as a defining sacrament in the life of secular individuals. As reformers sought to compile a coherent code of marital legislation
and restrict monastic orders to more stringent codes of celibacy, theologians—troubled by the anti-matrimonial doctrines of religious heretics—
actively elaborated more positive definitions of marriage and attempted to
arrive at a unified definition of the sacramental value of the conjugal bond.
Hence the stage was set for a convergence of religious and temporal values,
and one can only wonder how it is that medievalists have so long insisted
on emphasizing a dichotomous view of medieval attitudes toward secular
love.

Paradoxically, marriage reforms gave sharper definition to distinct features of the religious and secular vocations while inherently giving more immediate religious significance to what was merely the secular *civitas*, previously associated with innate moral corruption. Contemporaneous struggles against Cathars and against Nicolaism, the simultaneous defense of marriage here and insistence on chastity there, implied the existence of two distinct, if equally orthodox, sets of moral standards predicated in turn upon distinct social orders. At the center of these divisions, it was marriage practice (with all of its attendant, long-established, and often secular features—e.g., procreation, social mobility, courtship, love, sexuality, fidelity, etc.) that played a pivotal role, negative in relation to religious orders and positive in relation to secular orders. This meant that Church doctrine (anticipating a more comprehensive Thomistic synthesis) now attempted to embrace a widening sphere of autonomous secular values.[4] The status of Charlemagne, the crusades, and the investiture struggles are all related instances of a similar historical trend over the course of several centuries. In the reign of Charlemagne, for example, the papacy was forced to acknowledge the sovereign military and political might of the Frankish ruler but attempted to reconcile the secular power to its own providential claims by endowing the emperor with a divine mission. The crusades, likewise, involved the integration of secular martial values to an ostensibly religious mission. The investiture controversies revealed the tensions underlying the struggle for superiority between the two spheres of power (papal and secular) and gave even greater breadth to the notion of parallel autonomy. As both sides advocated their claims to ultimate authority and defended the importance of their respective prerogatives, new bodies of rational *apologia* emerged which deepened the philosophical divide between the realms of Church and State. Emerging marriage doctrine provided the Church with another avenue for recuperating and imposing a unified theological vision by incorporating the laity as a whole into a broader providential order, as third in a hierarchy in which Christian saints and martyrs came first. The sacramental theory of marriage, therefore, provided a new exegetical framework for the exploration of a pivotal secular institution with roots in pre-Christian mythology and literature. Nor was there any essential contradiction in this meeting of profane and religious traditions, to the extent that religious marriage doctrine clearly emphasized the uniquely secular role of the conjugal bond. Celtic and Classical texts available to twelfth-century authors, along with the mythological subtexts preserved in them, would now take on a fresh interest and provide traditional material rich in exemplary potential.

This new discursive space finds its counterpart in the unique social and vocational status of the secular court cleric and finds its fulfillment in a literary production intended for aristocratic patrons.[5] Not only did the nascent appreciation of Roman letters provide for new models of secular edification based on Classical models of *urbanitas* but marriage itself (in terms of both its amorous and political import as well as its new theologicial dimensions) provided a nexus for integrating secular and religious themes and values. This convergence of religious and secular itineraries would also prove to be a source of troubling ambiguity and a rich vein for potential (satiric) irony as courtly clerics explored the doctrinal, poetic, social, and metaphoric boundaries of so many competing models and discourses. In light of emerging sacramental doctrine and canonical controversies, the marriage theme would no doubt engage contemporary issues and unite Christian teaching with overtly secular concerns.

In any event, the preoccupation with love, marriage, courtship, and adultery in Old French narratives need not arise directly from the influence of lyrical genres. In addition, the attempt to define and apply notions of "courtly love" in a consistent and convincing manner will clearly fail as long as scholars draw indiscriminately on a mixed corpus of genres—genres born of distinct impulses and with distinct aims. If the definitions of medieval *love*, of *courtly love*, of *fin'amor* are both pivotal and problematic for our modern approaches to much of Old French secular literature, we need to acknowledge that such notions were not fixed terms but fluid elements in a broad area of religious, historical, and social speculation. Rather than assign them a fixed value, we would be better rewarded to understand the subtle and varied parameters of the debate. For this reason, we need to examine the deeper dimensions of the fundamental ideological conflict surrounding secular marriage *mores* as they converge with the evolving doctrinal glosses.

The concepts central to the new doctrine and which most influenced literary representations of marriage were those of marital affection and mutual spousal consent. In Duby's two competing models of medieval marriage (the "ecclesiastic" and the "lay" models) consent and affection are the key elements that distinguish the tenets of Church doctrine from those of secular feudal custom. In all the narratives making up our corpus, marital conflict has a critical if not central role, and the question of consent plays a pivotal role in each of these marriage plots. We will see, however, that the narratives do not reflect an author's perception of social reality or a finite historical setting. Instead, the authors juxtapose and contrast marriage practices based on the competing models. Throughout, the consent model prevails over the essentially coercive features of customary (aristocratic) marriage. Interestingly, the edifying logic of the narrative teleology does not take the form of an imposed Christian predication, but develops out of the course of secular events and conflicts and according to the ideological aims and biases of the individual author. Hence, our study of marriage fictions will not be limited to a catalogue of doctrinal and literary correspondences. Indeed, we will find that the twelfth-century *roman* exploits the marriage theme as a vehicle for critiquing a broader array of feudal conflicts to the extent that the consensual marriage bond functions according to the same logic as the feudal bonds that ratify and ensure the stability of larger social alliances.

Of course, the fundamental importance of marriage as a social institution reminds us that aristocratic marriage practice also entails critical historical and economic consequences. Indeed, the landmark marriage of Henry II Plantagenet with Eleanor of Aquitaine, the former wife of the French monarch Louis VII, would radically change the political configuration of Western Europe. Aristocratic marriages were themselves fraught with intrigue and conflict for these very reasons, and the narrative plots might be illuminated further as mirrors of actual events. Anthime Fourrier points to the correspondences between the marital intrigue in *Cligès* and the marital negotiations carried out as part of the political machinations engaging the courts of Friedrich Barbarossa, Manuel Comnena, and Louis VII in the years 1170–71.[6] One marriage alliance devised by Louis and Barbarossa involved the German Emperor's infant daughter Agnes (born 1170) and Louis's son Philippe (born 1165). Here we also find potential echoes in the depiction of Partonopeus's adventures in *Partonopeus de Blois*. For the young prince Philippe suffers a hunting mishap which could have inspired the depiction of Partonopeus's marvelous adventure in the forest. Likewise, the German Agnes will be courted in 1173–4 by the Egyptian sultan Saladin (who seeks to betroth the German princess to one of his sons) just as Partonopeus's primary rival in arms and love will be Margaris, a Persian sultan. *Girart de Roussillon* might contain echoes of the political relation between Girart of Mâcon and the German Emperor Friedrich Barbarossa, which is transformed from enmity into a new alliance by virtue of Barbarossa's marriage with Beatrice (1156). Beatrice is crowned queen of Burgundy in August 1178, shortly after Barbarossa's own coronation ceremony (Fourrier, *Courant réaliste*, 192; 200–1).[7] Likewise, the double betrothal depicted in the epic poem (between Girart and Elissent and between King Charles Martel and Berthe) along with the precipitous marital "swap" imposed by the king, might reflect the simultaneous betrothals in 1179 between Count Henri of Champagne and Isabelle of Hainaut, on the one hand, and Badouin VI of Hainaut and Marie of Champagne, on the other. Two years later (1181), when the betrothals of the young couples are to be reconfirmed, Isabelle is now engaged to Louis's son, the present king Philippe Auguste, and the Young Count Henri consequently finds himself coupled with Isabelle's sister Yolande (Fourrier 192–3).[8] The theme of sequestration central to *Erâcle* also has a historic resonance in light of the sequestration of two politically powerful women, Beatrice of Burgundy and Eleanor of Aquitaine.[9] Despite so many intriguing echoes, however, the narratives in our corpus often remain difficult to date with certainty, and the elaboration of imaginative settings and adventures suggests a thematic import not fully accounted for by historical events

themselves. It is in the context of larger literary and intellectual developments that we will discover the most fruitful correspondences between
"history" and "fiction."

Indeed, the political dimensions of aristocratic marriage practice as well
as the conflicts and intrigues that plagued marriage interests and feudal
society at large would no doubt make the exploration of marital customs
all the more compelling. As various notions of love, affection, and passion,
of *cupiditas* and *caritas* converge (notions drawn from secular lyric, from
Ovid, from Roman law and patristic tradition), we will also see subtle distinctions in the treatment of amorous *mores*. While adultery is attenuated
in the context of coercive marriage unions, desire itself remains a problematic and impulsive force. Different texts will orient the question of "love
madness" in novel ways. However, the new marital model becomes somewhat less morally ambiguous in later narratives. As the emerging doctrines
gain broader acceptance, sentimental refinement and esthetic beauty
become ethical values in their own right which govern the representation
of secular heroes as well as narrative structures and themes. In the end, we
will see that emerging views of love and marriage do not merely redefine a
narrow aspect of secular life, but lead to a reorientation of secular ideology and link the Old French *roman* to a larger western tradition.

NOTES

1. "The same inference which Kittredge draws from the rationalizing in French
romances, we are obliged to make from the extraordinary amount of conflation of
variants to be observed in our [the Modena] archivolt" (35).
2. The widely used term "courtly love" has come to encompass so many ill-defined
or divergent assumptions that critics would debate the very existence of the sentimentality it assumed while advocating in recent decades the very abolition of the
term itself (cf. Newman).
3. See Jean Seznec, *The Survival of the Pagan Gods* (1953) (Princeton: Princeton
University Press, 1972) [translation of *La Survivance des dieux antiques*, 1940].
4. It is of course true that twelfth-century thought does not yet refer to the notion
of a "double truth." In my analysis of the *Roman d'Enéas*, however, I identify
numerous and sophisticated categories of moral, philosophical, and historiographical judgment. These categories of thought clearly allowed twelfth-century authors
to view ostensibly profane texts according to multivalent levels of exemplary or
typological exegesis (see chapter two).
5. On the formation and outlook of a new class of clerics active in secular affairs,
see C. Stephen Jaeger, *The Origins of Courtliness: Civilizing Trends and the
Formation of Courtly Ideals 939–1210* (Philadelphia: U of Pennsylvania Press,
1985).
6. Anthime Fourrier, "Encore la chronologie des œuvres de Chrétien de Troyes,"
Bulletin Bibliographique de la société internationale Arthurienne 2, (1950): 70–88.

7. For additional documentation, see also Paul Fournier, *Le Royaume d'Arles et de Vienne* (1138–1378) (Paris: Picard, 1891).

8. Cf. H. D'Arbois de Jubainville, *Histoire des ducs et des comtes de Champagne* (t. 3) (Paris: Durand, 1861).

9. Eleanor was sequestered by her husband Henri II between 1174 and 1189. It seems that Beatrice was sequestered for political motives (*circa* 1150) by her uncle William (cf. Fournier, *op. cit.*, pp. 12 and 21).

Marriage Fictions in Old French Secular Narratives, 1170–1250

CHAPTER ONE

Marriage Fictions and the Meaning of "Courtly Love"

The study of amorous and marital intrigue in the twelfth-century vernacular narrative of France is both a compelling and a daunting prospect. It is compelling because medieval French and Occitan are the Romance vernaculars that give expression to the richest body of lyrical expression and romantic fiction in an age when tales of conjugal love dominate vernacular verse narratives.[1]

Immediately compelling for these reasons, this same study is equally daunting given the emergence, in twelfth-century Europe, of new forms of amorous discourse and matrimonial doctrine. Moreover, we cannot ignore the copious and often contradictory critical legacy surrounding the topic of love in medieval literature—a legacy that now proves burdensome, overwhelming at times. In 1968, with the publication of a collection of papers entitled *The Meaning of Courtly Love*,[2] the extent of the critical impasse had become abundantly clear. These papers demonstrated that virtually no agreement existed among the field's most eminent scholars with regard to the fundamental significance of *amour courtois* and the broad range of amorous tropes that it encompassed. At present, medievalists are still to some degree in the throes of this disconcerting critical "crisis." Like any crisis, however, it has led to new opportunity.

First, critics such as Benton (1968), Press (1970) and Paden (1975) have provided important, albeit negative, critiques of underlying assumptions about medieval love literature and medieval attitudes toward love. Secondly, some critics, such as Shirt (1982) and Gaunt ("Marginal Men") have begun to re-examine literary representations of amorous themes in light of new controversies. But just as the negative critiques provide a corrective but not an explanation, the new approaches often seem intent on reducing literary irony to the certainty of rigid and dogmatic prescriptions and fail to explain the preoccupation with subversive forms of love (in lyric and narrative works) in relation to prominent features of medieval moral doctrine and medieval social history.

Let's look briefly at Gaunt's study ("Marginal Men") of the poet Marcabru. First, Gaunt rejects the assertion that *fin'amor* represents an adulterous ethic which voices a reaction to emerging theological marriage doctrines. He first casts doubt on the historical approach by arguing that historians such as Duby and Brundage "betray a startling ignorance of the troubadours they cite to support their views" (55). Referring to Paden's survey of troubadour lyrics (a survey whose results demonstrate a rather negligible preoccupation with adultery), he notes that all of the poems cited by Paden which do talk about adultery date from the first half of the twelfth century (56). This was the period in which Gratian's edicts (*Decretum*, c. 1139) gave formal expression to the Church's condemnation of adultery (for both sexes) within the context of broader legal developments with regard to marital doctrine. [3] This period in history also witnesses a new willingness on the part of ecclesiastical authorities to actually *act* on such pronouncements (59). An accurate assessment of troubadour poems from this period does not, however, support the idea that medieval troubadours shared a common reactionary attitude. Rather, argues Gaunt, the preoccupation with adultery in the work of Marcabru, seems to take up the theological position within the context of a debate specific to his generation and to contemporary troubadours, a debate over "the rights and wrongs of adultery" (60–1). To conclude his argument, Gaunt asserts that Cercamon and Marcabru express a condemnation of adultery in their poetry. Yet, when we look closely at the examples cited by Gaunt, we are reminded of Press's admonitions with regard to the fundamental "inadequacy" of the troubadour corpus as a source of reliable and coherent information for any unified view of medieval attitudes toward love and marriage (328–9). Indeed, the subjective orientation of troubadour poetry makes it inherently problematic as a source for extra-literary pronouncements. What *moral* import of any kind can one attribute or derive from Cercamon's vehement condemnation of adulterous husbands, when, in the same breath, he reserves for himself the right to sleep with their wives (whom he makes beholden for any promised sexual favors)? It is a question whether the irony precludes any basic moral prescription. If it does give voice, as Gaunt argues, to a condemnation not of adultery but of promiscuity, it is hard to see what (common) moral ground the speaker pretends to claim for himself and his audience.

In the case of Marcabru, it is equally difficult to discern beneath the sometimes prophetic tenor of his warnings the moral ground on which he stands. In one example cited by Gaunt, for example, Marcabru emphatically lashes out against promiscuity. But does Marcabru's animosity echo a theological perspective, or does it not in fact reflect a rather vaguely defined secular ethic based on notions of honor and moderation? Gaunt ignores this question altogether, despite the fact that it appears crucial in some way to the larger historical claims and logic of his argument. Instead, he invokes the rather cryptic poem as evidence of a conscious moral rule and prescription: "Nowhere in this poem is there any hint that Marcabru permits a woman one lover but no more" (62). This "reading" of the poem's import is based on the moral prescription ascribed to Cercamon's thoroughly ironic *Ab lo pascor*: it supposedly answers Cercamon's limited proscription of adultery with a totally inclusive one. Indeed, Gaunt seems to confuse a rhetorical orientation with a moral one when he says that "Marcabru's condemnation of promiscuous women . . . sets him squarely in a tradition of Christian moralizing" (63). The ideological basis for this condemnation has really nothing to do with theological notions of *sin*, but rather with the principle of legitimate descent central to the legitimacy of aristocratic prerogatives (63)—a concern which Gaunt's own observations reveal to be associated not with the emerging theological doctrine of marital consent, but with a lay model of marriage which privileges parental control over betrothals, endogamy, and repudiation at will (57–9; cf. Duby *Two Models* 7–12; 68–9). Gaunt points out, however, that the condemnation of adultery for *both* sexes (as opposed to woman only—which served in the lay model to ensure purity of bloodline) was unique to the Church model and also a feature of Marcabru's poetry (58–9, 63). Again, however, Marcabru's key terms suggest a purely secular ethic based on *mesura* and which, not unlike Cercamon, praises love while condemning promiscuity:

> Mesura es de gen parlar,
> E cortesia es d'amar;
> E qui non vol esser mespres
> De tota vilania is gar,
> D'escarnir e de folleiar,
> Puois sera savis ab qu'el pes.
> C'aissi pot savis hom reignar,
> E bona dompna meillurar;
> Mas cella qu'en pren dos ni tres
> E per un non si vol fiar,
> Ben deu sos pretz asordeiar
> E sa valors a chascun mes.[4]

> [*Moderation is from courteous speech and courtliness from loving, and the one who wishes to remain without blame must shun all base manners, all deceit, and all careless indulgence. Then he will be wise, as long as he thinks about it. For in this way may a wise man rule, and a virtuous woman improve. But she who takes two or three, wishing not to be faithful to only one, surely does she taint her reputation and her merit forever more.*]

In speaking of *mesura*, *amor*, and *cortesia*, Marcabru does not seem to express any notion that extra-marital relations are either by nature, or by reference to theological concepts of flesh and sin, inherently impermissible. Nor is it clear to me how in light of this poem's vague reference to the rule of moderation and its explicit prescription of amorous fidelity (. . . *E per un non si vol fiar*), it is to be cited as evidence that Marcabru condemns adultery because he provides "no rider that a woman is allowed one lover" as Cercamon allows (Gaunt "Marginal Men" 63). If it is a condemnation of adultery, it in no way appeals to any specific Church authority or religious rules. In his study of western sexuality, Michel Foucault has demonstrated that classical discourse on sexuality and ethics which has its roots in Greek philosophy typically a) addresses aristocratic male behavior in particular, b) excludes any fundamental concept of carnality and sin, and c) finds its key rationale in the Greek concept of moderation. While it may not (could hardly be) a continuation of Greek ideals in their particular, Marcabru's moralistic stance would seem to evoke the kind of secular (and aristocratic) criteria for an ethic of sexual conduct that Foucault has abstracted from his own survey of Greek aristocratic traditions. In this schema, sexual conduct provides a field of ethical conduct where prescriptive norms are of little concern compared to individual demonstrations of superior self-control, natural virtue (exemplary *bienséance*), and moderation—this last principle being reflected in the troubadour notion of *mesura*, and excluding not only excessive lust, but also unnatural abstinence. In speaking of the moral principles underlying the Greek notion of ethical conduct, Foucault states:

> Elle [cette reflexion morale] est une élaboration de la conduite masculine faite du point de vue des hommes et pour donner forme à leur conduite. Mieux encore: elle ne s'adresse pas aux hommes à propos de conduites qui pourraient relever de quelques interdits reconnus par tous et solennellement rappelés dans les codes, les coutumes ou les prescriptions religieuses. Elle s'adresse à eux à propos des conduites où justement ils ont à faire usage de leur droit, de leur pouvoir, de leur autorité et de leur liberté: dans les pratiques de plaisirs qui ne sont pas condamnées, dans une vie de mariage [où?] aucune règle ni coutume n'empêche l'homme d'avoir des rapports sexuels extra-conjugaux, dans des rapports avec les garçons, qui, au moins dans certaines limites, sont admis, courants et même valorisés. Il faut comprendre ces thèmes de l'austérité sexuelle, non comme une traduction ou un commentaire de prohibitions profondes et essentielles, mais comme élaboration et stylisation d'une activité dans l'exercice de son pouvoir et la pratique de sa liberté. (29–30)

> [*It [this moral consideration] is a theory of male behavior elaborated by men and intended to shape their actions. Moreover, it does not have anything to do with those practices affecting men and which derive from certain interdictions recognized by all and solemnly addressed in any number of religious codes, customs, or prescriptions. It concerns men with regard to precisely those practices through which they exercise their rights, their power, their authority, and their liberty: in the practice of pleasures which are not condemned, in a conjugal life [in which] no rule or custom pre-*

vents the man from having extra-marital sexual relations; in relations with boys, which, at least within certain limits, are permissible, rather common, and even esteemed. It is necessary to understand these concerns with sexual austerity not as an echo of or commentary on serious and essential prohibitions, but as a deliberately stylized practice carried out as an expression of male power and liberty.]

If Marcabru's interest in adultery is (to some degree or another) indebted to contemporary marriage controversies, it is nonetheless difficult to impose one moral-historical discourse upon another, to approach one poetic discourse and ethic from another genre of socio-historical discourse, from an ethical discourse with its own semi-distinct genesis. Likewise, if Marcabru's poetry cannot accurately be said to advocate an ethic of adulterous love that emerges as a reaction to Church proscriptions, it is equally difficult to situate it historically as an extension of Church pronouncements and concepts. Indeed, to make sense of Marcabru in the context of contemporary theological teaching on marriage, one must either ignore the secular orientation of his ideology or redefine the historian's most basic assumptions about the antagonism between lay and ecclesiastical models of marriage. This brief view of Marcabru's position within the context of the Church's marriage reform efforts, reminds us that the doctrinal leaps made by Church thinkers during the twelfth century reflect only a beginning in a long process, a beginning in which Church policies represented not a monolithic dogma but the first uncertain steps to create almost *ex nihilo* a new domain of sacramental theory (Noonan 1973 425; *Dict. de theologie cath.* 9 (2): col. 2068; 2109; 2125–6). Throughout this period, theological innovations remain tied to immediate pragmatic concerns and were designed not to radically displace, but rather to "infiltrate" and "absorb" lay practices built upon well-established local customs (Duby *Two Models* 17–18; Colish 2: 2125–26).

Duby himself provides an alternate to his model of the opposition between lay and ecclesiastical practices in order to account for dynamic views within secular practices themselves. For Duby, the troubadour views on marriage and adultery are not to be referenced according to emerging theological doctrines as Gaunt argues, but to an ideological division within lay (aristocratic) society based on the opposition between young bachelors (*juvenes*) and married property owners (*seniores*) (Duby *Love* 30–1; *Two Models* 107–8).[5] Given the hybrid cultural matrix out of which new marriage doctrines emerge, however, all of Duby's reductive dichotomies need to be applied with some caution (cf. Colish 2: 658, and 658 note 478). With regard to the relation between literature and these different social and ideological regimes, Duby tends, again, to be rather reductive in his formulations. He asserts that literary discourse was "subject to pressure from two opposing models, the ecclesiastical and the secular, each of which prevailed, depending on which literary genre was involved" (*Knight* 211). In our study of marriage fictions, we need to keep in mind that the contemporary literary discourse is indeed open to both distinct and potentially exclusive ideological regimes, but that twelfth-century culture (in the area of marriage customs at least) also offered the possibility for broadly syncretic forms of ideological play and dialogue.[6]

Another example of an ostensibly innovative approach that falls prey to its own reductive view of contemporary social history and its relation to literary invention is David Shirt's article "*Cligès*—A Twelfth-Century Matrimonial Case-Book?" (1982). In his article, Shirt refers to twelfth-century legal doctrines to provide a fresh approach to medieval marriage fictions, and he provides a convincing analysis of subtle parallels between Chrétien's fictional marriage plot and (sometimes rather localized) points of contemporary law. In essence, Shirt shows that Chrétien (who "picks his way through the twists and turns of the matrimonial maze with the sort of skill and assurance which would impress many divorce lawyers today") goes to great lengths to place, in terms of legal definitions, the affair and eventual marriage between Cligès and Fénice outside the sphere of legal and moral transgression (84–5). In doing so, Shirt necessarily reorients long-standing views about the work's reactionary views on adulterous love (cf. Maddox "Trends" 732–3). For Shirt, *Cligès* neither condemns nor condones adultery, but celebrates marriage by "legitimizing the conduct of Cligès and Fénice from the point of view of secular and canon law" in a way that "establishes harmony and knit between the main body of the narrative of the second part of his romance and its epithalamatic conclusion" (85). However, if (as Shirt claims) Chrétien merely "hoodwink[ed] generations of modern critics into believing that *Cligès* is his attempt to come to terms with the 'unacceptable face' of troubadour morality" (85), we are left asking why so much burlesque irony befalls the illicit schemes of the star-crossed lovers who follow in the footsteps of the (in)famous Tristan and Iseut about whose reputation (in the Middle Ages) it is difficult to draw any positive conclusions (cf. Maddox "Trends" 732–3; Le Gentil "Tristan" 126; Fourrier 40–1).

These unanswered questions point to a more crucial flaw in Shirt's general claims—a flaw that characterizes the larger critical impasse with regard to medieval marriage fictions. Specifically, Shirt clearly implies that contemporary marriage doctrine debates provide the principal key to Chrétien's elaborate marital fiction. In so doing, Shirt narrowly links literary fiction to social reality. We feel the stress and strain of this reductive leap when Shirt asks us to believe that "Chrétien's courtly public would have . . . no difficulty in appreciating the legal side of his romance" (85), while he simultaneously rejects the likelihood that this same public might be capable of appreciating the literary (i.e. "the *Tristan* romances with all their ramifications") or diplomatic ("the machinations of the Franco-Byzantine intrigue") intrigues that critics have referenced in previous studies of the tale (76). Shirt's final rationale for such a sweeping claim is based on his reference to the important role of marriage contracts in the lives of Chrétien's princely patrons. Yet when have aristocratic families *not* been preoccupied with marital engagements and intrigue, and how do such mundane preoccupations account for Chrétien's convoluted, fanciful and ofttimes burlesque plot?

Although Shirt effectively demonstrates that *Cligès* is a drama concerned at least as much if not more with marriage doctrine than it is with subversive passion and adultery *per se*, his approach exaggerates the link between contemporary reality and literary fiction. Of course, Shirt is correct in pointing out that "the period 1150–1200 was crucial in the formation and development of matrimonial legislative practices in Western Christendom" (76), but important developments in marriage doctrine can equally be said to span the period 1050–1150.[7] This broad historical coincidence between emerging marriage doctrine and fictional marriage tales cannot go unnoticed, however, and it no doubt holds more promise for new critical approaches than some earlier sociological theories in the quest for (what Shirt calls) a "critical justification" for popular narrative themes in social history.

In the 1960's, for example, the eminent medievalist Erich Köhler adopted a different kind of sociological approach. Making reference to Frappier's insistence on the intimate and dynamic relations between "le fait social" on the one hand and "le fait littéraire" on the other, between "la réalité historique" and "l'image poétisée," Köhler ("Observations") was inspired to construct a socio-economic interpretation of Provençal lyric. Köhler sees the courtly love ethos as a purely poetic ideal which arises as a response (*la projection sublimée*) to the frustrated economic and social aspirations of a nascent social stratum of lower nobility (*la basse noblesse*) (28). Arguing that an essential link exists between the new lyric motifs and the poet's relative social inferiority (35), Köhler is of course hardly able to account for the fact—a fact he himself cannot fail to acknowledge (37)—that Duke William IX of Aquitaine, the earliest known practitioner of the lyric forms to become popular in twelfth-century southern France, is a figure of superior social rank. Furthermore, William's poetry is often quite explicit in its erotic interests and hardly seems representative of a lyricism based on metaphorical deferral. No doubt the individual circumstances of some singers may inspire such metaphorical tropes, but this would seem to reflect the subjective nature of the genre and the diversity of its practitioners more than any intrinsic relation between secular love lyric and economic class. It is one thing to observe the transfer of customary and verbal forms from one domain to another, but it is much more to define a poetic genre, in terms of both genesis and function, as a medium of collective equilibirium within a broadly-defined cultural milieu and as a response to an equally broad notion of real economic scarcity and competition (as Köhler seeks to do) (33–4). In the end, Köhler's attempt to accommodate

William and his *oeuvre* to a socio-economic model by postulating that high-ranking lords were "forced" into poetic conformity in order to secure the enduring fidelity of vassals seems rather far-fetched (38).[8] In fact, whereas Köhler suggested that troubadour poetry served as a tool of collective secular cohesion (34), other critics who seek to explain courtly love in terms of social reality have in fact constructed either an inverse interpretation of the literary tradition or a nearly inverse model of social attitudes. For Georges Duby and Moshé Lazar, for example, the sentiments of courtly love express real (not sublimated) desires, as a means of seduction (Duby) or as an expression of immediate carnal desire (Lazar).

Duby saw courtly love as a literary and social "game" or "ruse" which operated as a strategy of seduction and which developed as a response to a specific socio-economic transformation: namely the increased population of young, unmarried knights who were in the quest of well-endowed brides (*Knight* 92–4). "Love" (erotic desire) was seen as a poetic corollary for predatory desires rooted in the economic alienation suffered by disenfranchised bachelors whose willful prerogatives undermined the stability of the conjugal bond (40 and 225). Ultimately, however, various contradictions belie the problematic relation between literature and reality. Despite Duby's valuable insights into the development of medieval marriage practices, his writings provide no consistent claims with regard to the way in which literature conforms to reality in this context. On the one hand, secular love literature is said to reflect a destabilizing form of concupiscent desire which stands in opposition to the attempts by Church leaders to define a stable form of marital affection based on sacramental and charitable love (i.e. *amor* vs. *caritas*) (Duby *Two Models* 13–14; 55). On the other hand, Duby sees courtly love sentiments as a direct extension of clerical misogyny (*Knight* 218: both love poets and religious writers "put forth a male morality based on a primal fear of women and a determination to treat them as objects").

Whereas Köhler sees the amorous ethic as a subliminal projection that enhances the status of the (desired) woman and compensates for economic alienation, Duby tends to see courtly love as a patriarchal strategy that makes women into victims of deception and pawns for ambitious but disenfranchised knights. Lazar, on the other hand, argues that troubadour lyric should be taken at face value in its expression of passionate carnal desire: "La discussion porte sur la question de savoir s'il s'agit d'un adultère purement spirituel, d'un jeu littéraire, ou bien (ce que la plupart des médiévistes n'osent et ne veulent envisager) d'un amour adultère tout court" (63). Like other critics, however, Lazar makes distorted and overly reductive claims (54: "la *fin'amors* adultère est une conception commune à tous les troubadours sans exception"), and his "literal" reading of the corpus seems to preempt any thought of discriminating between lived experience and attitudes on the one hand, and metaphoric tropes on the other:

"la *fin'amors* n'est pas spirituelle ni mystique. Elle est avant tout le désir d'une union physique avec l'aimée, elle évolue en fonction de la récompense espérée par le troubadour" (85). In Lazar's view these precepts represent an ethos that can in no way be reconciled either with the institution of marriage or with Christian morality (12–13, 60–61). Moreover, Lazar supposes that underlying this poetry is a vast current of medieval ideology constituting an "amoral" ethic ("nettement opposée à la morale chrétienne") with its own rules and systematic precepts (12, 137, 148). Lazar therefore moves easily from the lyric to the narrative corpus by relying on reference to this common cultural-historical matrix. Thus Thomas's *Tristan* is to be appreciated according to a controversy over the relative merits of free love (*amour libre*) and conjugal love: "L'amour triomphe complètement de la société et de la morale chrétienne. La morale de la *fin'amors* a raison de toutes les objections" (161, cf. 172).

Although our focus is on representations of marriage (and adultery) in narrative works, we see that broad categories of critical thought and approach have, without much hesitation, moved between literary (lyric and narrative) and social (historical and economic) domains. The very format of Lazar's work, with its introductory chapters on *fin'amor* and its additional chapters on narrative works, provides an illustration of the way in which medieval tales of love and marriage are likely to be glossed in accordance with lyric models—models that often comprise not only formal tropes or ornamentation but are claimed to give expression to contemporary amorous mores.

TRISTAN ET ISEUT: VERSION COMMUNE AND VERSION COURTOISE?

When, however, with decades of *Tristan* scholarship at her disposal, Sarah Kay makes reference to the "elusive" nature of textual "point of view" in the *Tristan* poems (187, 195), we are warned against viewing medieval textuality as a vehicle for transparent (secular or religious) sermonizing. Indeed, the Old French *Tristan* poems have generated an astonishing variety of critical responses—responses at times diametrically opposed. Frappier ("Structure" 257) indirectly identifies the source of this potential ambiguity in pointing to the pagan-Celtic origins of the twelfth-century French *romans* and to the alien aspect of its native motifs. For Frappier the legendary narrative material constitutes the core of what he calls the *version commune* of the *Tristan* legend. The supernatural, fatalistic and irrational aspects of the pagan saga would have proved, in Frappier's estimation, ill suited to a literary milieu infused with the kind of lyric models and refined sentimentality emanating from southern court circles ("Structure" 259: "Ce premier état du *Tristan* . . . ne répondait guère et même s'opposait à l'idéal nouveau de courtoisie et d'amour courtois dont la vogue, venue du Midi, se répandait de plus en plus durant la seconde moitié du

XIIe siècle, dans les milieux mondains d'oïl"). This inherent contradiction is purportedly resolved with the emergence of Thomas's version, the *version courtoise*. For Frappier, virtually every aspect of Thomas's adaptation of the narrative material can be seen as a reflection of the poet's apparent familiarity with the conventions of *fin'amor* ("Structure" 259). Perhaps because other critics were not so bold in attempting to force the ethically problematic features of the tale into a clear and definitive ideological mold, and perhaps because the tale seemed to form a natural extension of the apparently adulterous ethos of Provençal lyric, Frappier's view gained exceptional currency and certainly served to accommodate an important medieval narrative tradition to an emerging critical theory of courtly love. Let us examine a few essential points relative to Frappier's view of the representation of love, adultery and marriage in Thomas's text. Does Thomas's version reflect lyric models and conventions? Does it celebrate an amorous ethic which consents to adulterous transgression?

Frappier's argument relies heavily on his interpretation of Thomas's handling of the love-potion motif. Rendered a purely symbolic emblem of *Amor*, Thomas's *philtre*, says Frappier, "ne sert plus . . . à excuser le péché des amants. Ceux-ci n'ont maintenant nul besoin d'un alibi moral. L'obéissance à l'Amour est devenu leur seul devoir. Leur seul devoir et leur volonté unique" ("Structure" 273). Fourrier attempts to make a similar argument as to the "symbolic" nature of the love-drink, but his own remarks remain rather obtuse with regard to the exact nature of the love in question (70–1: "le signe ésotérique de l'amour"). In a footnote, however, Fourrier backpedals somewhat when he acknowledges the limits of Thomas's fictional sleight of hand and the enduring emphasis on the fatalistic nature of the love drink (71, note 258: "Néanmoins, il subsiste aussi chez lui [Thomas] des traces et comme un souvenir de la tradition qu'il atténue . . ."). Frappier wants us to accept that *Amour* retains its autonomy, although the lovers are intended to appear free from servitude to any fatal passion such that the evident "permanence" of the potion's effect should not be taken as a reflection of its "fatalistic" import (274–5). Frappier ultimately formulates the supposedly decisive opposition between the *version commune* and Thomas's "courtly" reworking in rather tautological terms: "A une fatalité sans option a succédé l'option d'une fatalité" (273).

Had Thomas sought to emphasize the rational and ennobling aspects of love, it seems he would have attenuated the fatalistic aspect of the love attraction much further. Indeed, it is hard not to take note of the passive posture and unrewarded suffering experienced by the lovers, and numerous critics do not share Frappier's view. Even Fourrier's analyses, which prefigure and support Frappier's with regard to the influence of a courtly love ethic in Thomas's version, lead to different conclusions. First of all, Fourrier points out that the love theme is essentially tragic and constitutes an *esthetic* drama not an ethical posture (13; 107).[9] In highlighting the *trag-*

ic and *dramatic* aspects of Thomas's narrative, Fourrier's own response to the text departs from Frappier's with regard to the concept of free rational choice which for Frappier is an essential component of the troubadour ethos (Frappier, "Structure" 265). Frappier will argue that Thomas attenuates the literal import of the fatalistic love potion, but while Fourrier also defined the potion as a native motif which was "rationalized" by being made into a mere esthetic "symbol" of love, he sees love represented as a tragic and all-consuming passion whose supernatural dimensions are only mildly attenuated (107: "Plus que l'hymne de l'amour courtois, le *Tristan* de Thomas en est le drame. Ses enchantements opèrent encore").[10]

It is ironic, however, that Thomas seems to level the field of amorous heroism by bringing Mark into the same magical fold and by making him a victim of the *philtre* along with Tristan and Iseut (Fourrier 71). Frappier has difficulty with regard to Thomas's treatment of Mark, and he contradicts his own assessment of the potion's relative significance as he attempts to negotiate between the text and his thesis:

> Il convient surtout d'observer que l'innovation de Thomas, malheureuse ou non, est commandée par l'idéologie de la *fine amor*, toujours prompt à immoler le personnage et les droits du mari. Le détour imaginé par le remanieur anglo-normand avait pour but de rendre le roi Marc victime de la fatalité, et non plus les amants. (276)

> [*It is worth noting that Thomas's innovation, ill conceived or not, is governed by the ideology of courtly love—ever ready to sacrifice the person and the rights of the husband. The variation imagined by the Anglo-Norman teller of the tale was designed to make King Mark the victim of the fatal power, rather than the lovers.*]

Fourrier will argue that this is Thomas's way of accounting for the ostensibly unexpected way in which Mark consistently fails to punish firmly the adulterous lovers—at least he seems to require more than a little nudging from his council of peers. But this seems to trivialize the importance of Thomas's creative modifications. Indeed, as Fourrier also points out, Thomas's version tends overall to enhance Mark's stature (i. e. in contrast to the *version commune*) as an historical figure. He depicts Mark as the supreme monarch of England.[11] In this regard, the narrative text adds a political dimension and motivation typically absent from lyric models. Thomas evokes contemporary political affairs by using the lure of a united Anglo-Irish realm as a motivation to force Mark to forge a marital alliance (despite his pledge to Tristan) (63–4). Thus Mark's role seems to engage the action in a novel way; he is more than a mere foil for the celebrated figure of the adulterous lover.[12] Here amorous ethics play a key but nonetheless subordinate role in relation to a larger feudal drama—a drama where marriage prerogatives based on political or personal expediency and external rites come into conflict with those based on passion, clandestine pledges and fatalistic mysteries. Fourrier's own remarks add to this set of observations by revealing that Thomas also emphasizes (more than the *version commune*) Mark's prior verbal engagement to Tristan as well as the inner suffering of not only Tristan and Iseut, but of Mark and Iseut aux Blanches Mains (63; 103–4). Therefore, Thomas seems not to have been thinking predominantly of the archetype of the amorous triangle but was interested in exploring the tragic dimensions of the *Tristan* myth as the basis for a mystical counterpoint to secular ideology. Troubadour lyric, while it may speak to the superior merits of the lyric subject, does not provide such a context for defining amorous rivalries. We will see that in narrative works, on the other hand, marriage in all of its dimensions political and amorous, lay and religious, social and providential, provides a central motif for an exploration of a metaphoric union or nexus between the spheres of the secular and the divine.

Other critics articulate quite differently what they discern to be the general ideological import of Thomas's version. Le Gentil ("Tristan") brings into relief the less edifying aspects of the lovers' comportment in Thomas's version. Referring to Thomas's two lovers as the "victims" of a fatal passion, he says: "le conflit dont elles [les victimes] souffrent surgit et s'achève en catastrophe . . . Telle étant en dernière analyse la signification du mythe de Tristan, il est clair que Thomas, en se réclamant de la courtoisie, s'est mépris sur le thème qu'il traitait, ou en a fâcheusement restraint la portée" (119–20). Likewise, P. Jonin defines Thomas's posture *vis-à-vis* courtly love conventions in terms that are almost diametrically opposed to those put forth by Frappier:

Les troubadours donnent aux amants le pouvoir d'orienter leur vie amoureuse. Les auteurs de Tristan ignorent ou bafouent cette autonomie du sentiment. Aussi, l'amour d'Iseut et de Tristan n'est-il pas courtois dans toute la mesure du moins où sa naissance et son développement échappent au jeu habituel des passions humaines (181).

[*The troubadours give lovers the power to orient their own amorous life. The authors of the* Tristan *story ignore or deride this freedom of sentiment. Likewise, the love of Tristan and Isolde is not courtly love, at least to the extent that its birth and development escape the typical play of human passions.*]

These divergent viewpoints not only raise questions about the actual cogency of Frappier's thesis, but they also cast doubt on the assumption that narratives based on an adulterous affair can quickly and easily be resolved according to the tenets of *fin'amor*. Le Gentil observed evident discrepancies between lyric models of adulterous love and the kind of desire and triangular relations that obtain in Thomas's *Tristan et Iseut* ("Tristan" 121–2). Le Gentil points out that Tristan suffers feelings of carnal separation and abstinence and not the pains and pleasure of amorous desire as such. Likewise, Tristan's attempt to assuage his suffering by indulging in a purely carnal union is not a response that conforms to lyric models. Nor do periods of separation from Iseut seem to inspire Tristan with delicate and ennobling sentiments. Thomas's version provides the following description of Tristan's state of mind at the moment the hero is contemplating marriage to Iseut aux Blanches Mains:

A sa dolur, a sa gravance
Volt Tristrans dunc quere venjance;
A sun mal quert tel vengement
Dunt il doblera sun turment;
De paine se volt delivrer,
Si ne se fait fors encombrer;
Il en quida delit aveir
Quant il ne puet de sun voleir. (vv. 214–221)

[*To his own grief and chagrin does Tristan thus willfully seek vengeance; to his own harm does he seek such revenge by which he will double his own sufferings; wishing to free himself from pain, he only adds to it. In this way he thought to gain some satisfaction when he could not have his way.*]

The exceptional psychological penetration inflecting Thomas's narrative was seen by Jonin as potential evidence of lyric influences. Jonin, however, points out that *stylistically* Thomas's narrative and his narrative interjections do *not* conform to the stylistic devices typically encountered in the twelfth-century *romans* (316). Jonin points out that Thomas's work reveals, instead, an unmistakable predilection for orthodox sermonizing:

"A moins qu'il ne se livre à de laborieux essais d'analyse psychologique, Thomas nous oriente avec une nette préférence vers les généralités morales. Ses remarques sont presque toujours longues et lourdes et la monotonie de leur ton diminue encore leur intérêt. On ne saurait donc voir une contribution courtoise dans les commentaires pesants dont Thomas accompagne l'évolution sentimentale d'Iseut et de son amant" (316).

[*Unless he is absorbed by arduous elaborations of psychological analysis, Thomas has a clear preference for directing us toward moral generalities. His remarks are almost always long and dense, and their monotonous tone renders them even less appealing. Thus, it is hard to see any influence of courtly love in the dense commentaries which Thomas appends to the sentimental evolution of Isolde and her lover.*]

The tone of Thomas's own commentary on Tristan's decision to seek solace in a cynical and deceptive marriage clearly provides an example of the narrative style and intellectual posture described by Jonin:

Pur le nun e pur la belté
Que Tristrans i ad trové
Chiet en desir e en voleir
Que la meschine volt aveir.
Oez merveilluse aventure,

Cum genz sunt d'estrange nature,
Que en nul lieu ne sunt estable!
De nature sunt si changable,
Lor mal us ne poent laissier,
Mais le buen puent changer. [etc.] (vv. 230–239ff.)

[*On account of the name and the beauty that Tristan finds in her, he is seized by desire, wishing to have the young maiden. Now you will hear something marvelous and learn that people are of a strange nature, never finding peace of mind! They are by nature so inconstant they can never abandon their base habits but can deviate readily from virtue.*]

It is no doubt this kind of moral analysis that led Le Gentil to question the extent to which lyric models contribute to Thomas's vision. With regard to Tristan's marriage, Le Gentil says of Thomas: "Il a mis en pleine lumière la cruauté à la fois gratuite et injuste que le triomphe final de la *vertu* courtoise devait inévitablement avoir pour rançon" ("Tristan" 122). Contrary to Frappier, Le Gentil sees in Thomas's version a penetrating and critical insight into the deceptive and ambivalent aspects of love and adultery:

Ainsi l'amour même, et surtout l'amour adultère, se juge à la générosité des actes qu'il est capable d'inspirer. Que cette générosité vienne à disparaître, devoirs et obligations extérieurs à l'amour reprennent toute leur signification, parce qu'il n'y a plus d'amour au sens noble, au sens courtois du terme. Le sentiment coupable . . . redevient, faute de prouesse et d'héroïsme, d'élégance et de formes, ce qu'il est pour la loi sociale et religieuse, un crime qui mérite châtiment. ("Tristan" 124)[13]

[*Thus love itself, and above all adulterous love, is judged by the acts of generosity it is capable of inspiring. Should this generosity fade away, external obligations once again take on significance because there is no longer any ennobling love in the courtly sense. Without prowess and heroism, without elegance and form, the guilty desire . . . once again becomes what it is according to social and religious law, a crime deserving of punishment.*]

Jonin notes that Thomas's moralizing posture is particularly flagrant in the extended 130-line commentary on the parallel marital and amorous trials that beset Mark, Iseut, Tristan and Iseut aux Blanches Mains (316). It is also important to point out that Thomas's description of this four-way marital conundrum fails to reflect the lyric model of the adulterous triangle. In fact, Thomas shows little conscious interest in the motif of adulterous love as such. It is properly speaking 1) the mutual trials and tribulations of passionate desire and 2) the problem of love (absent or present) in marriage that seem to preoccupy the Anglo-Norman writer. We have seen how Mark's partaking of the potion allied him with Tristan and Iseut; likewise, Tristan himself is placed on a nearly parallel plane with Mark by virtue of his own marital dissatisfactions. In other words, Frappier's influential thesis never resolves the essential discrepancies that exist between the courtship and seduction strategies that motivate the lyric *canso*, on the one hand, and the archetypal conflicts at the mythic center of the Celtic Tristan tale. Thomas elaborates with quadrilinear equilibrium a drama of amorous suffering which forms an ironic counterpoint to the self-aggrandizing perspective of the lyric subject who wants to be at the center of the conventional adulterous triangle:

> Entre ces quatre ot estrange amor:
> Tut en ourent painne e dolur,
> E un e autre en tristur vit;
> E nuls d'aus nen i a deduit. (vv. 71–4)
>
> [*Between these four was a strange love: all of them suffered pain and distress; both the one and the other lived in sadness, and none of them found in love any pleasure.*]

Just as Frappier deemed Tristan's adulterous liaison with Iseut to be modeled in Thomas's telling of the legend on a positive concept of *fin'amor*, critics commonly assume that tales of marriage in the twelfth-century French *roman* derive equally from lyric models. To account for the discrepancies between a lyric tradition that ostensibly celebrates adultery and a narrative tradition that frequently integrates love and marriage, they divide medieval conceptions of courtly love into two distinct but genetically related branches, the southern and northern . According to this perspective, the preponderance of plots culminating in marriage represent not a contradiction with regard to the tenets of *fin'amor* but a slightly more "conservative" moral posture and an attempt at *reconciling* the precepts of *fin'amor* to a model of conjugal union based on mutual affection.[14]

CLIGÈS

Yet, when we look at the marital subplots that serve as introductions to the primary plots of Thomas's *Tristan et Iseut* and Chrétien's *Cligès* we find no evidence of an inherent tension between love and marriage. The Norwegian translation of Thomas's version presents no trace of any unresolved contradiction when it states that Kanelangres consummates his marriage to Blensinbil "conformément à la loi et au sacrement du mariage" immediately upon returning to Brittany and introducing her to his court as his *bien-aimée* (524). Likewise, marriage and love appear to exist simultaneously in Thomas's subplot. Blensinbil appears no less afflicted by the pain of love after her marriage than she did before the marriage (524–5). Chrétien portrays the marital drama of Alexander and Soredamors in quite similar terms. Neither the queen, nor Alexander, nor Soredamors seems to perceive any inherent contradiction between amorous sentiment and betrothal. In his study of love and marriage in *Cligès*, Noble argues that regardless of the marital motif, the depiction of love itself departs in this text from troubadour models: love is not controlled by reason, but is an impetus, rather, to irrational behavior (41). Therefore, if marriage provides a *solution* to the violent or irrational impulses of love, the tension that may exist between the desire and the institutionalized sacrament does not reflect troubadour depictions of either love or spousal indifference. Instead, the depiction of love would seem to reflect, or at least derive from, Roman and Christian concepts of irrational passion and find its relation to marital sacrament within the same cultural construct.[15] This compels us to revisit Köhler's assertion (based on a distinction between "northern" and "southern" ideologies) that marriage provided (for northern thinkers) the only viable and virtuous outcome for a concept of ennobling love: "Ainsi se posait la tâche de faire concorder la nature éducative de l'amour courtois et les exigences d'une relation amoureuse conduisant au mariage, c'est-à-dire de prouver que la tension nécessaire au perfectionnement et à l'ennoblissement de l'homme existait également dans l'union amoureuse légale" (*Idéal et réalité* 64). Unfortunately, Chrétien provides no clear expression of his adherence to such a programmatic purpose. However, Chrétien's depiction of the dynamic relation between love, innate merit and political harmony provides some positive basis for Köhler's argument that Chrétien sees love and marriage in terms of the necessary compatibility between natural law and social order. The queen frames the union of Alexander and Soredamors (a union that will serve as a prescriptive reference point for the more complex relation between Fénice and Cligès) with an incisive statement on the importance of amorous restraint (vv. 2284–7), and Alys's improper and illegitimate usurpation of Cligès's role as emperor is thematically linked to the rivalry over Fénice by parallel references to the natural laws of individual perfection which inform a natural (as opposed to arbitrary and absolutist) order of political and amorous superiority (vv. 2464–2528; 2687–2779). In the initial confrontation between Cligès and Alys, this gap between contingent circumstances on the one hand and what the terminology of natural reason and beauty suggests to be on par with the designs of providential dispensation on the other is made manifest in the corresponding gap between Alys's lack of real feudal legitimacy and his cynical effort to retain the nominal title of emperor:

> Or ot Alys, se il ne fet
> A son frere resnable plet,
> Que tuit si bani [baron] li faudront,
> Et dit que ja plet ne movront
> Qu'il ne face par avenant;
> Mes il met an son covenant
> Que la corone li remaigne,
> Comant que li afeires praigne.
> Por feire ferme pes estable
> Alys par un suen conestable
> Mande Alixandre qu'a lui veigne
> Et tote la terre mainteigne,

Mes que tant li face d'enor
Qu'il ait le non d'empereor
Et la corone avoec li lest ; . . . (vv. 2531–45)

[*Now Alis understands that should he fail to come to a reasonable agreement with his brother, then all of his armed supporters will abandon him.
He states that whatever pact they put forth, he will in turn agree on the
condition that he be allowed to keep the crown, whatever may happen. To
ensure a lasting peace, Alis has one of his constables request Alexander
come to him and assume possession of all the land but grant him enough
good will as to allow him to retain the imperial title and the crown.*]

In this narrative content, the same *mystery* that informs amorous fatality
and individual destiny governs both the marital and political fortunes of
the protagonist. Hence the tale's preoccupation with adulterous conflict
betrays less the influence of the troubadour *canso* than it does the dynamics of a Celtic or otherwise Indo-European sovereignty and sovereignty
bride myth. Likewise, Köhler's formulation fails to emphasize the inverse
dynamic by which publicly sanctioned marriages, like that of Alys to
Fénice, can also lead to instability. Once again it is marriage and (as we
shall see in chapter three) its twelfth-century doctrinal modalities, not
amor, which provide for the dynamic *trope*, and troubadour lyric provides
virtually no dynamic model of marriage.

As in Thomas's *Tristan*, secretive love becomes an important motif and
appears to pose a threat to the well-being of the lovers (e.g. vv. 2280–3).
According to the conventions of *fin'amor*, love is seen as a sentiment that
confers distinction upon the lover, but the lover is often motivated to conceal his sentiments, ostensibly because the love is adulterous or the desired
relation is, socially, illicit. In the Alexander-Soredamors subplot there are
no such motivations for the impulse toward secrecy. Soredamors's haughty
bearing leads her to scorn her amorous sentiments. Love is concealed not
because of an illicit relationship, but because sentimentality is deemed foolish. As with Soredamors, so too do Cligès and (even more so) Fénice experience shame and dishonor in conjunction with amorous desire.[16] It is, in
fact, hard to find reference to any single and clear motivation for the lovers'
refusal to confront the treacherous Alys and make public their mutual sentiments.[17] This suggests that secrecy serves Chrétien both as a narrative
expediency and as a motif related to the deferral of conjugal bliss, as a
potential *cause for* (not symptom of) adulterous transgression (cf. Hanning
22–3). This assessment correlates with the work's psychological depictions
which set in relief, as Noble has pointed out, the modesty and timidity of
the lovers (40). Secrecy functions, therefore, not so much in relation to
external society or social customs but with regard to the mutual and frank
expression of amorous sentiments between the two lovers:

"Alixandre," fet la reïne,
"Amors est pire que haïne,

Qui son ami grieve et confont.
Amant ne sevent que il font
Qant li uns vers l'autre se cuevre. . . .
De ce trop folemant ovrez
Que chascuns son panser ne dit,
Qu'au celer li uns l'autre ocit:
D'Amors omecide serez. . . ." (vv. 2263–7, 2282–5)

[*"Alexander," said the queen, "love is worse than hate in the case of some-
one who harms and bewilders his beloved. Lovers do not know what they
are doing when the one conceals things from the other . . . With this you
get so carried away, that neither one of you speaks his or her mind.
Concealing your thoughts, you kill your beloved: you will be a Love
homicide!"*]

As the queen continues, she expresses an acute insight into the naive
aspects of amorous sentiment, namely the difficulty of "taking the first
step" and getting beyond purely subjective ruminations:

En amor a molt greveuse oevre;
Et molt torne a confondemant;
Qui ne comance hardemant
A poinne an puet venir a chief.
L'en dit que il n'i a si grief
Au trespasser come le suel. (vv. 2268–73)

[*In love there is much strenuous labor, and many things go awry. The one
who does not enter into it with determination will rarely make it to the
end. It is said that there is nothing more difficult than the first step.*]

While Chrétien's psychological shadings undoubtedly reflect a rich literary inheritance including not only troubadour models but also, perhaps
primarily, Ovidian ones (to which Chrétien makes explicit allusions), the
specific features of the Alixandre-Soredamors subplot do not reflect any
clear parallels between models of courtly love in which socio-economic
conditions, or an adulterous ethos, play a defining role. Marital transgression serves as a dialectic counterpoint to the marital idyll staged initially in
the union of Soredamors and Alexander, and the problem of age is joined
to the psychology of amorous reticence in a way that allows Chrétien to
explore the surprises that inevitably arise in conjunction with new definitions of marriage and new laws concerning the consummation of the marital bond.[18]

In social and vaguely ideological terms, marriage (again according to the
queen's admonitions) should provide an institutional structure for both the
harmonious progression of love and for its assimilation into the public
sphere. This seems, at least, to be the immediate if perhaps not the ultimate
significance of the queen's slightly cryptic but rather definitive statement on
this matter:

> Or vos lo que ja ne querez
> Force ne volanté d'amor.
> Par mariage et par enor
> Vos antranconpaigniez ansanble.
> Ensi porra, si com moi sanble,
> Vostre amors longuemant durer. (vv. 2286–91)

> [*Hence I advise that you never seek willful or coercive love. Better you should live together in marriage and in honor. Only this way, as it seems to me, will your love be able to endure.*]

This foretale to the main action of *Cligès* does, therefore, set in relief the theme of marriage, but it is not the only important theme to be raised nor does the story prepare us for what Shirt has defined as a narrative inspired by an interest in contemporary canon law controversies *per se*. Both Thomas and Chrétien elaborate similar marital intrigues that explore the deceptions and irrational consequences associated with concealed love. Moreover, both authors depict an essentially identical concept of *amor* inspired by Ovidian tradition—a tradition "in which desire (*cupido*) and suffering (*passio*) played leading roles" (Baldwin 137). Defined accordingly and distanced from the problematic associations with *fin'amor* and "courtly love," the essential distinction between Thomas and Chrétien also emerges with more immediate clarity. Thomas "glosses" the tragic vision of the *Tristan* legend according to the operations of amorous suffering. The outcome is not a moral statement on love; rather, the tragic outcome is accounted for by an amorous psychology (here Augustinian *concupiscentia* and Ovidian *cupido* intersect) that depicts love and courtship as governed by emotional suffering and perpetual disillusionment and which distances itself in voyeuristic fashion from the tragic marvel it scrutinizes. In the end this distance is expressed through a didactic posture and thinly veiled (if veiled at all) admonishment:

> Tumas fine ci sun escrit:
> A tuz amanz saluz i dit,
> As pensis e as amerus,
> As emvius, as desirus,
> As enveisiez e as purvers,
> (A tuz cels) ki orunt ces vers. . . .
> Pur essample issi ai fait
> Pur l'estorie embelir,
> Que as amanz deive plaisir,
> E que par lieus poissent troveir
> Choses u se puissent recorder:
> Aveir em poissent grant confort,
> Encuntre change, encuntre tort,
> Encuntre paine, encuntre dolur,
> Encuntre tuiz engins d'amur! (482; vv. 38ff.)

[*Here Thomas finishes his tale—one dedicated to all lovers, to the pensive and the infatuated, to the jealous, to the ones filled with desires, to the perverse, to everyone who will hear these lines . . . As a lesson he has written this, to embellish the story such that it might be pleasing to lovers, that they might be able to discover here or there things that they themselves have experienced, that they might draw from it comfort in the face of change, in the face of wrongs, in the face of pain, in the face of sorrow, in the face of all the deceptions and intrigues of love!*]

Baldwin contends that while the theological tradition approached desire through the vocabulary of sin, "the Ovidians approached desire with ambivalence" (137). Here that ambivalence, which informs the logic of the palinode at the end of Andreas Capellanus's famous treatise, reveals itself as a veritable topos for the convergence of a vernacular, secular and heroic depiction of love, on the one hand, and an implicit clerical and moral discourse on the other. Thomas, far from celebrating a transcendent amorous ethos, continually depicts love (both in and out of wedlock) as a source of emotional anguish, internal contradictions and tragic suffering. Chrétien, on the other hand, tends to see love as a manifestation of a natural order conforming to reason and natural law, whose ambivalence stems only from the *potential* and subjectively imagined gap between private desire and public honor. When channeled into a publicly disclosed marriage, however, love does not give way to ongoing pain, deception, and dissension, but comes to an indefinite and happy stasis. It is the tension, not between adulterous desire and conjugal bonds, but between private volition and public harmony,[19] between one course of historical destiny and another, that seems to drive the narrative of *Cligès*. Before asking how the text's elaborate intertextual dynamics shed further light on Chrétien's central concerns, let us consider another medieval narrative that seems to draw attention to the conventions of *fin'amor* in its portrayal of latent marital rivalry within a framework of epic conflict.

GIRART DE ROUSSILLON

If the marriage fictions we have just studied show little direct debt to lyric models and social realities, it would seem unlikely to look for any such influence in a twelfth-century *chanson de geste* whose marital intrigues are largely eclipsed by epic motifs (rebellious baron, an unsympathetic view of royal privilege, an unjust king, failed negotiations, vendettas and internecine strife). At the same time, however, *Girart de Roussillon* seems unusually susceptible to a reading based on troubadour influences. Not only does Girart stand at the head of a territorial alliance that privileges France's *pays d'oc* (Combarieu du Grès and Gouiran 8), but the very language of the poem, although a dialectical composite, contains many Occitan elements.[20] The editors of our edition, referring to the unknown poet as "troubadour plus que trouvère" (6) refer to Pfister's erudite conclusions on the poem's composite language:

> Il situe la langue du *Girart de Roussillon* original dans la région de Vienne, mais elle se serait moulée dans une *scripta* littéraire, dont le noyau serait constitué par les éléments occitans, et qui étendrait son influence sur le Lyonnais, le Dauphiné, le Forez, mais aussi sur la Provence, l'Auvergne et le Velay. Enfin, M. Pfister explique le caractère composite de la chanson par la rencontre de trois composantes: "la langue spontanée employée par l'auteur original de la région de Vienne, la langue artistique de la poésie des troubadours et la langue épique de l'ancien français."[21]

> [*He associates the original language of the* Girart de Roussillon *with the region of Vienne, but suggests that this language has been molded into a literary* scripta, *the roots of which would be made up of Occitan elements and whose influence would extend to such regions as Lyon, the Dauphiné, Forey, and also Provence, Auvergne, and Velay. In brief, Pfister explains the composite nature of the poem as resulting from the encounter of the constituent elements: the immediate language used by the original author from the region of Vienne, the artistic language of the troubadours, and the Old French language of epic poetry.*]

René Louis has also insisted upon the poem's aggregate form and is tempted to see the extant epic as the object of successive accretions at the hand of more than one author.[22] Louis's claims are reinforced by the fact, among others, that the poem seems to move—and sometimes quite abruptly—between different stylistic and generic registers. In short, the conflict between Girart and the king is framed by a "prologue" that recounts a marital intrigue infused with romantic motifs and sentiments (approx. 597 vv.) and by an "epilogue" (2,650 lines) providing a rather spiritual denouement that develops Christian themes and expresses religious ideals.[23] Indeed, were we to adhere to Louis's speculations, we would be tempted to attribute the poem's prologue to the work of an unidentified "renouveleur" of southern origin (Poitevin) active in the court of Eleanor of Aquitaine (Louis 3: 76) and to see it as the reflection of direct "troubadour" influences (2: 347–8) with the amorous drama subjecting Girart and Elissent to the "conditions morales de l'*amor de lonh*" (2: 403–4). These speculations mesh with Louis's opinion that love is a central theme in the work (2: 416) and that the poet portrays love as being incompatible with both marriage and carnal desire (2: 416: "l'amour vertu, opposé aussi bien à la passion coupable qu'à l'union conjugale").[24]

When we recall the main outline of the marital subplot at the center of the poem's so-called prologue, it becomes easy to see why Louis saw the poem as expressing an anti-matrimonial bias. The prologue (the first 597 lines of a work amounting to just over 10,000 lines) relates the details of an amorous rivalry comprising two acts of apparent marital transgression committed by King Charles and Count Girart respectively. We learn that according to a prior accord involving King Charles of France and the Emperor of Byzantium and designed to help the Emperor build a more effective alliance for the defense of Christian territories against pagan incursions, the Emperor's two daughters (Berthe and her younger sister Elissent) are to be betrothed respectively to King Charles and to his most renowned vassal Girart, Count of Roussillon, heir to the Duchy of Burgundy. After a largely idyllic description of the military collaboration of Charles and Girart in a victory over a host of pagan aggressors, Girart—having demonstrated his valor—travels to Constantinople where he and the (absent) king are solemnly betrothed to their respective future brides according to formal rites presided over by the Pope and a number of wise dignitaries. The absent king (who expresses apprehension over Girart's growing power and fame) suddenly burns to obtain news of the two brides however, and to know which is the more beautiful. Learning that Girart's fiancée (Elissent) outstrips her sister in physical charms, the jealous king rushes to meet the party as it returns to France from Constantinople so as immediately to claim Elissent (i. e. the younger daughter who is designated as Girart's fiancée) for his own. The dignitaries and all the parties are shocked by the king's action, and Girart finds his honor deeply offended. The Pope, concerned about the need for a strong imperial alliance and the welfare of larger Christendom, eagerly tries to arrive at a new settlement in order to obviate an embarrassing repudiation. In the end, Girart reluctantly resigns himself to Charles's will and agrees to marry the elder sister instead of Elissent. At the same time, however, Girart remains tied to Elissent by explicit bonds of affection even though she is now destined to become the queen of France. Hence Charles's initial transgression against marital protocol is succeeded by an antagonistic rivalry over Elissent. First, Charles imposes his will in such a way that Elissent becomes a mere pawn in the contest over sovereignty that erupts between a king and a powerful vassal. Secondly, Girart and Elissent are prevented from marrying, and as a consequence their otherwise legitimate bond takes on subversive aspects. Girart's reference to Elissent as *midonne* and his stated intent to serve as her legal champion should the need arise (vv. 527-9) combined with Elissent's own admission of love ("Que plus l'aim ke mon paire ne mon seinor," v. 586) help to account for Louis's impression that the prologue reflects the conventions of *fin'amor*.

However, the notion that love and marriage are incompatible *per se* is hardly supported by the marital plot. First of all, we have seen that if Girart and Elissent love out of wedlock it is by no choice of their own; but rather it is as the outcome of the king's capricious imposition. Secondly, Louis's own impression of the opposition between love and marriage seems to depend upon a questionable interpretation of a single line in which the poet, according to Louis, describes Girart's marriage with Berthe as one forever absent of amorous affection:

E Girarz sa muller ac esposade.
Con plus la tec lo cons, mais l'a amade. (vv. 536-7)

While the editors of the most recent bilingual edition of the poem (based on the Oxford manuscript version) have translated these lines: "Une fois Girart marié, son amour pour celle qu'il avait épousée n'allait cesser de grandir avec le temps" (Combarieu du Grès and Gouiran), Louis (who apparently reads *mais* as "never" 2: 36) arrived at a perfectly antithetical reading ("qu'il n'aimera jamais d'amour). Louis's reading is indefensible, and while we can see how the poem might have prompted his interpretation, we will soon see that the marital subplot really does not reflect the lyric tropes or adulterous ideals associated with *fin'amor*.

Let us return briefly to Girart's use of the term *midonne* with reference to Elissent and to Elissent's own bold expression of adulterous love. Several critics, Le Gentil (1957), Hackett (1982), Combarieu du Grès (1986) have already commented on the discrepancy between the treatment of the love theme in *Girart de Roussillon* and contemporary conceptions of troubadour love. Combarieu du Grès ("Elissent" 27–8) points out that Elissent, by virtue of the king's preference, is now elevated to the rank not of countess but of queen of France. Accordingly, she has come to rank as Girart's superior just prior to the very moment that he refers to her as *midonne* (v. 528) and only shortly before the poet makes reference to Charles's jealousy. William Paden points out, however, that critics of troubadour lyric have ascribed a falsely restrictive meaning to the use of this term. Paden's careful study of a large troubadour corpus leads him to conclude that "*domna* in Provençal . . . is simply a form of polite address or reference. It establishes no feudal relation between the poet and his lady; moreover, it connotes nothing at all regarding the lady's marital state" (33). The same conclusion applies to the term *midons*: "Often *midons* and *domna* seem to be interchangeable in the lyric. Neither word necessarily shows that the lady to whom it is applied has higher social rank than the speaker, much less that she is married" (36). Nor, according to Combarieu du Grès ("Elissent"), does the appearance of the term here prove sufficient to show that the poet's portrayal of the marital intrigue derives from the conventions of courtly lyric: "Ces quelques traits ne suffisent pas à déterminer la relation amoureuse courtoise dans sa spécificité. Les divergences sont nombreuses. . . . Il n'y a, de la part de Girart, ni poème en l'honneur d'Elissent . . . ni prouesse de sa part dédiée à la même intention . . ." (28).

Le Gentil ("*Girart*") concurs. He argues that the mutual affection between Girart and Elissent is in no way tainted by illicit intent: "Les liens qui se sont noués jadis n'étaient pas ceux de l'amour courtois. . . . Les gages qu'ils ont échangés l'ont été devant témoins. . . . Si les sentiments éprouvés sont complexes, ils sont purs; ils n'ont pas à se cacher et personne ne les suspecte, parce qu'ils n'ont jamais donné prise au soupçon" (478–9). The ambiguous status of this affectionate bond—both extramarital and "pure" (a love defined rather vaguely by Louis as *amour vertu*)—seems less problematic when we realize that the *gages* referred to here by Le Gentil actually represent a semi-clandestine exchange of marital vows.

In events leading up to this pivotal scene, Girart had already "tested" Elissent's sincerity, and Elissent had affirmed her love for Girart before two witnesses ("Cui volet melz, donzele, mei o cest rei? —Se Deus m'ajut, charz saigne, eu am plus tei" vv. 463–4). Gaunt has argued that Girart's question, posed as it is, reveals that the marital rivalry is merely a metaphor of sorts for the political and personal rivalry between vassal and king, and that the apparently amorous relation between Girart and Elissent is in fact devoid of love with Elissent being nothing more than a token object in a "homosocial" contest between two men:

> Le rôle des femmes dans *Girart* ressemble donc à celui qu'elles ont dans la poésie des troubadours et, à l'instar de la dame d'une chanson troubadouresque, Elissent et Berthe sont marginalisées pour que le poète privilégie une relation masculine dont elles ne seraient que les médiatrices. Charles et Girart, les deux héros, seraient donc liés par le désir *homosocial,* terme qui désigne l'utilisation par un homme hétérosexuel d'une femme pour entamer une relation avec un autre homme. ("Femmes" 305)
>
> [*The role of the women characters in* Girart *is similar therefore to the role of women in troubadour poetry. Akin to the lady of the troubadour tradition, Elissent and Berthe are marginalized so that the poet might foreground a masculine relationship for which the woman plays only a mediating role. Thus the two heroes, Charles and Girart, are linked by* homosocial *desire—a term which designates a heterosexual man's use of a woman as a means of engaging a relation with another man.*]

However, the ties of mutual affection between the two are underscored by their dramatic and sentimental reunion following Girart's exile (epilogue, vv. 7823ff.) and more immediately (in the prologue) by the fact that Girart's final decision to go along with the king's imposed bride exchange and to marry Berthe is depicted as an act of reciprocal fidelity toward Elissent ("Eu prendrai ta seror per amor tei" v. 468). This need not mean that the marital contest is entirely divorced from the political struggle. On the contrary, as the feudal conflict brings into the focus competing perceptions as to the extent and legitimacy of Charles's political sovereignty, we must keep in mind what the marriage drama in the poem's "prologue" reveals: namely, that Elissent (now queen of France) stands like Fénice not merely as the pawn in a masculine rivalry but as an emblem of sovereign destiny and legitimacy and thus, problematically, as the the object of Girart's real and affective fidelity.

As for the apparently illicit love affair between Girart and Elissent, it becomes clear that the action is not modeled on concepts of adulterous love, but is built upon an interest in love and marriage. Girart's relation to Elissent certainly provokes Charles's jealousy, and Gaunt's ("Femmes") evocation of "homosocial" theories is perhaps relevant with regard to the conflation of amorous and political rancor that besets the king. Girart's situation, however, involves a perhaps more rare phenomenon. The text clearly implies that he is in fact married to two brides: he is nominally married to Berthe (although he will later develop an affectionate bond with her as well, cf. vv. 536–7 discussed above), and he is bound by both affection and formal vows to Elissent. This assessment of Girart's ambiguous status is supported by the poet's depiction of the quasi-clandestine ritual which unites Elissent to her former fiancé. Elissent makes a formal pledge of love (with God and a small audience of onlookers called as witnesses), and the ring she gives as a gift to Girart provides a symbol of the lovers' union and of their sworn fidelity. Finally, Elissent pledges to Girart the most valuable part of her dowry:

> Bertolais e Gervais, qu'es riu contor,
> Vos m'en siaz ostage e lui autor,
> E vos, ma char[e] seur, ma tan fesor,
> E en apres Jesu lo redemptor,
> Que doin per ist anel au duc m'amor;
> E doins li de mon oscle l'aurieflor,
> Que plus l'aim ke mon paire ne mon seinor. (vv. 580–586)

> [*You Bertolais be my witness in this matter, and you Gervais the scribe.*
> *You, my dear sister, my confidante along with Jesus the savior knowing*
> *that with this ring I give my love and I pledge the flower of my bridal gift*
> *to the duke, whom I love more than my own husband and lord.*]

Thus we see that the rivalry between Charles and Girart does not involve a triangular model of adulterous courtship. On the contrary, the double bride motif should remind us instead of the Celtic tale type that figures in the genesis of the *Tristan* legend and reappears in two other twelfth-century Old French narratives—the "Eliduc" attributed to Marie de France and *Ille et Galeron* by Gautier d'Arras.[25] Moreover, *Girart de Roussillon* elaborates a complex marital drama that actually highlights competing models of marriage, i.e. ones based on external authority and political interest, as opposed to those based on mutual affection, spousal consent and individual autonomy.[26] We also see that the author has established at least a minimal figurative link between the amorous intrigue of the prologue and the conventional epic (i. e. military and political) conflict that follows to the extent that the marital ambiguity and rivalry parallel a similar ambiguity that emerges with regard to the contest over sovereignty. Once again, as in *Cligès*, we find a marital plot whose complications involve not an adulterous ethos, nor a metaphorical struggle for dominion, but a literal power struggle whose underlying mythic profile involves a ritual bride contest and on to which have been grafted parallel concepts relating to both marital and feudal concepts of fidelity, affection and legitimacy. In chapters three and four we will attempt to identify and explore the intersecting theological and social ideologies that twelfth-century authors exploited as a means of infusing an enduring archaic myth with a new and compelling wisdom.[27]

PARTONOPEUS DE BLOIS

In evoking the sovereignty motif along with its dynamic concept of the relation between the hero's bride and the hero's political and economic fortunes, we find ourselves in some way having come full circle with regard to historical approaches to "courtly love." We recall, for example, that Köhler and Duby both saw "courtly love" and its cultivated ethos of secular, amorous courtship as an imaginary *ersatz* that would compensate for (Köhler) or serve as part of a duplicitous response to (Duby) the economic disenfranchisement suffered by a broad population of lower nobility. For Köhler courtly love was a hypergamous fantasy that compensated for economic inferiority. For Duby, the courtly love ethos represented a fundamentally cynical, misogynist strategy of seduction that served the interests of disenfranchised knights seeking a hypergamous conquest (*Two Models* 107–8; *Love* 14–15). Likewise, when Duby defines "courtly love" as an elaborate and duplicitous courtship stratagem, he too is alluding to narrative, not lyric, texts (Duby *Two Models* 107–9). Indeed, if Provençal lyric provided a language of courtship and affection, the Celtic sovereignty motif independently provided a dynamic perspective on the relation between courtship and personal destiny—a dynamic relation that transcended the purely subjective fixations or fantasies of the lyric voice. Yet here too, Celtic and Provençal motifs reveal an uncanny affinity, for Breton lore (to judge by the *Lais* of Marie de France) provided a ready vehicle for marriage fictions based on marvelous fantasy and erotic desire in its elaboration of sovereignty motifs. Is it possible that attempts to seek troubadour models as a basis for marriage and adultery plots derived from Celtic models has led to a distorted view of prominent Old French narratives? Does the presence of Celtic sovereignty motifs nonetheless provide a basis for the socioeconomic influences defined by Köhler and Duby?[28] It would seem, for example, that the hero of *Partonopeus de Blois* is not unlike the figure Lanval in terms of his secret liaison with Mélior and the land of bliss over which this fairy figure rules. Given the images of lavish wealth associated with Mélior's kingdom and the initial cynicism of Partonopeus' initial amorous advances (his taking of Mélior by force), the tale's elaborate marital plot would seem to reflect the kind of socio-economic conditions and strategies emphasized by Köhler and Duby.

At first glance there is certainly some ambiguity concerning the social status of Partonopeus and his counterparts in the Old French *roman*— Tristan, Cligès, and Girart. On the one hand, all are involved in a conflict of one form or another with a ruling sovereign. On the other hand, these heroes are neither politically nor economically disenfranchised in real terms. Indeed, their conflict with a sovereign seems predicated on a virtual claim to sovereignty and on an exceptional birthright or noble patina.[29] Even in the case of *Partonopeus de Blois*, where we have a heightened sense of romantic fantasy and nostalgic longing, the main hero is not depicted as

an alienated and disenfranchised knight. In conformity with the archetypal scenario in *Tristan* and in *Cligès*, Partonopeus is nephew to the king of
France, and he actually rejects a cynical marriage ploy involving immediate
sexual and economic gratification when his uncle (the king) and mother
conspire to lure him away from Mélior with offers of money, land, castles
and a sensuous bride. Thus Partonopeus's quest, whatever it represents, is
not primarily a quest for a hypergamous conquest as a solution to economic and political alienation. Nor would Partonopeus appear to be motivated by purely material interests or content to achieve his ends through
cynical stratagems.[30] Thus, while various political and economic gains may
accompany the hero's marital conquest and may persist as vestiges of an
archaic sovereignty motif, the juxtaposition of competing marriage models
by the *Partonopeus* poet does not seem intended to isolate an exclusive link
between economic restoration and a marvelous bride figure. This observation leaves open the question of the poem's underlying ideological impetus,
and the nature of the larger correpondance that exists between the two
realms of marital intrigue.

It is interesting to note, therefore, that the *Partonopeus*-poet's treatment
of the love theme does not immediately serve to isolate one marriage from
the other. The archetypal potion that serves in the *Tristan* legend as an
emblem of uncompromising love and desire appears in *Partonopeus* not in
conjunction with the lover's attraction for a *marvelous* bride but as a device
in the parentally arranged betrothal. This casts a new light on the love-
potion which in the tales of *Tristan* and *Cligès* seems to reflect a kind of
magic fatalism to the extent that it is directly associated with the privileged
couple. In this case, however, it threatens to separate Partonopeus from his
elected partner, and we are forced to recall Thomas's version in which King
Mark also imbibes the fatal aphrodisiac. Thus the love potion serves
neither to make love into a *symbolic* entity (as Frappier and Fourrier have
argued with regard to Thomas, for example) or an idealized sentiment distinct from base desires. Rather, it reveals the poet's appreciation (crude or
not) of the subjective psychology and immediate physiology of desire and
evokes an image, not of Celtic magic, but of Ovid's own pseudo-scientific
approach to the same topic and of his warnings against the use of overpowering stimulants.

Yet, even to the extent that Partonopeus finds himself smitten as a
consequence of the potion's effect, the arousal of the hero's passion is in no
way lessened psychologically either by the lack of any amorous obstacle
(e. g. a jealous husband or rival), or by the prospect of marriage and unhin-
dered gratification. The woman who collaborates in the plot to seduce
Partonopeus makes marriage a pre-condition of their amorous and carnal
union: "Bien soutils hom seroit sopris / En tel liu et de tel pucele; / Car trop
est sage et pros et bele. / Cele respont qu'el n'ert s'amie / S'il nel espouse u
nel afie; / Et il en est si alumés / Que faire en violt ses volentés" (vv.

4014–20). Once again, therefore, we find a narrative where the protagonists reflect no hesitation when speaking of marriage in the context of
amorous desire. At the same time, the medieval poet seems capable of
understanding not competing forms of love (e. g. *cupiditas* vs. *caritas*) but
the powerful nature of love within the sphere of the individual psyche.

Moreover, Partonopeus's nobility does not depend on any essential doctrine of ennobling courtship and stems from his birthright: he is the son of
Clovis's sister Lucrece, and his ascendancy can ultimately be traced back to
Trojan nobility (vv. 365ff., 433ff., 535ff.). We also know that his inner perfection antedates his passion for Mélior: for it is Mélior who actually
"courts" Partonopeus, and it is the young knight's exceptional nobility that
determines Mélior's preference (vv.1346ff.). Love is, however, associated
with courtesy, nobility and refinement generally. For example, we learn
that love quickens Partonopeus during combat (vv. 3393–400) and that
love has an edifying virtue:

> Ensi set amors ensegnier
> Cascun home de son mestier:
> Cevalier de cevalerie,
> Et clerc d'amender se clergie . . .
> Cortesie done et largece. (vv. 3415–20)

> [*Thus love knows how to instruct each man in his vocation: the knight he
> perfects in chivalry, and the cleric in religion . . . he inspires courtesy and
> generosity.*]

Here again, however, the amorous intrigue departs from the more programmatic concepts often associated with *fin'amor*. This ennobling passion
does not become effective through a deferred union; rather, it coincides
with the consummation of desire. Only the actual marriage is deferred, but
this is not due to any inherent antagonism between love and marriage. The
poet, obviously interested in the edifying power of love, has therefore sal-
vaged this particular lyric trope by recourse to the fidelity "test" that
Mélior imposes on Partonopeus (i. e. instead of real sexual abstinence, the
hero is only prohibited from revealing his beloved's identity and from having any visual enjoyment of her body up until the appointed marriage
date). But marriage itself is portrayed as the ideal culmination of amorous
affection, and the tale's central concern is not with competing models of
love but with competing models of marriage and draws a distinction
between arranged or imposed betrothal and marriage born out of mutual
affection and consent.

MARRIAGE FICTIONS AND SOCIAL REALISM

It is clear that the unusual preoccupation with marriage plots in twelfth-
century narratives does not grow out of the conventions governing troubadour lyric. Both the paradigms governing the plots we have examined as

well as the contingent themes, often pseudo-historical and pseudo-political in nature, find only occasional parallels in the boasts, desires, and aristocratic ideals expressed in troubadour lyric. There is also scant historical evidence to suggest that the narrative plots we have examined are an immediate reflection of contemporary social experience.

In his rather exhaustive study of realism in the medieval *romans*, Anthime Fourrier scrutinized a number of French narrative works from our present corpus. Rarely, however, do his investigations lead him to see the primary aspects of fictional marriage plots as owing a debt to contemporary reality. Of course it is obvious that Thomas's *Tristan* perpetuates relatively archaic pagan traditions, and Fourrier's analysis reveals that only superficial aspects of Mark's portrait seem intended to confer a certain historical verisimilitude upon the otherwise marvelous events (44–5). Fourrier defines Thomas's "realism" almost exclusively in terms of "psychological realism" and not in terms of socio-historical realism as such (106–7). Ironically, this new sentimentality itself is expressed through purely literary tropes (cf. Faral). And while these tropes may be incorporated into narrative works, they are likely to prove unreliable indices to a given work's fundamental inspiration.

If we stop and recall how the amorous intrigues in *Cligès*, *Girart de Roussillon*, and *Partonopeus de Blois* contain broad parallels with certain archetypes transmitted in the *Tristan* narratives, we are led to conclude that whatever contemporary circumstances may have led to the "composition" of these tales, it is clear that the marriage fictions themselves have antecedents in distant traditions. Although Fourrier judges *Cligès* to be Chrétien's most "credible" work in terms of its relatively restrained use of the marvelous (154–6), he simultaneously acknowledges the role played by Celtic motifs at the most fundamental levels of novelistic conception: "le merveilleux féerique . . . lui offrait non seulement des ressources commodes pour agencer son intrigue, inventer des situations singulières et donner à ses héros un théâtre à leur taille, mais encore toute la poésie qui subtilement émane du mystère des vieux mythes" (115). Indeed, Fourrier's most compelling and insightful points of close analysis reveal the extent to which the plot of *Cligès* consists of a re-disposition of elements already present in Thomas's *Tristan* (152: "Tout, jusqu'à ses invraisemblances, s'explique par l'imitation de Thomas . . ."). While the marvelous subtext also serves in Fourrier's judgment as a counterpoint to the work's urbane wit and realistic touches, Fourrier shows these touches to consist primarily of descriptive details.

In those cases where historical parallels retain a potential relevance, they provide only a veneer of realism, echoes of contemporary social ideology but not an object of direct imitation. There may indeed be, as Fourrier has suggested, an echo of contemporary circumstances with regard to Anglo-Norman and Irish affairs in Thomas's *Tristan* (45). Moreover, the biography of Duke William V of Aquitaine (Count III of Poitou, late tenth century) may have contributed to the portrait of Count Fouque in *Girart de Roussillon* (Colliot). In fact, the historical duke's biography provided a rather dramatic tale of captivity, love and marriage (Colliot 14). Ironically, however, *Girart de Roussillon* only carries this historical memory in its most utterly distilled manifestation, so that the episode survives as the most marginal and fairy-like event of the whole epic, namely the tale of love between Fouque and the enchantress Aupais.

This lack of realism is all the more striking when we consider how readily the marital plots could be elaborated in historical terms. With reference to *Cligès*, Fourrier argues that beneath the archetypal conflict between the king and his young nephew we can detect echoes of contemporary marital strategies involving the most powerful aristocratic families of France, Germany, and Constantinople (161–173). But Fourrier's remarks also remind us just how different such high-stakes diplomatic maneuvers must have been in "real life" from the kind of sentimental and heroic rivalries operating in Chrétien's narrative. In his study *Love and Marriage in Chrétien de Troyes*, Peter Noble equally acknowledges that generally speaking, "romance, the genre which concentrates above all on love, particularly the awakening of love, makes little pretense of being realistic" (6).

In the end, "history" seems to provide only the most fluid of models for literary imitation in the medieval *romans* we have begun to examine. The year 1156, for example, provides a historical basis for a renewed controversy over the sovereign control of Burgundy (Fourrier 192), just as the year 1181 provides a historical example of a marital coup on the part of the sovereign Philippe-Auguste who usurps the bride (Isabelle of Hainaut) of the deceased Henri of Champagne (Fourrier 203)—a marital usurpation not too unlike Louis' forceful appropriation of Elissent in *Girart de Roussillon*. It would seem, however, that the medieval poet of *Girart de Roussillon* exploited historical echoes not as a source for the contours of his plot, but in order to add urgency and a recognizable field of ethical relevance to his own typological and reformist fiction. In this regard, *Girart de Roussillon* parallels *Cligès* in its preoccupation with the relation between love, marriage and social harmony. Religious reform efforts and emerging marriage doctrines assigned a pivotal role to the conjugal bond and defined marriage as the primary sacramental union designed to bring the secular orders into conformity with a divine plan for social peace (Duby *Knight* 215; 1978 17–18; Colish 2: 637–8).[31]

However, this preoccupation with marriage must have found, and been fueled by, unmistakable reverberations in literary traditions as well. For the same age that saw marriage as the quintessential sacrament of the secular individual also saw itself in the shadow of classical cultures founded upon the deeds of epic heroes such as Paris and Æneas whose amorous adventures determined the rise and fall of nations.[32] Indeed, Maddox has shown how, of all of Chrétien's romances, *Cligès* stands out for its concern with historical verisimilitude ("Discourse" 11). But Maddox also points out that this verisimilitude does not reflect any real historical pretense (as is proved by the fantastic events surrounding Fénice's stratagem and the lovers' adulterous liaison) as much as it serves to structure the dynamics of reception and define the work's ethical truth (1982 14–15). By bathing Fénice's irrational scheme and the lovers' adulterous affair in a marvelous atmosphere, Chrétien would seem to enhance his audience's appreciation of the intransigence of external (public) reality over against the interior world of love and subjective fantasy. Maddox's analysis also highlights the extent to which Chrétien's use of mimetic discourse adds historical dimensions to the work's ethical themes:

> Along with a kind of spatial mimesis unparalleled anywhere else in Chrétien, we find that temporality is used to heighten the referential illusion. In contrast with exclusive use of the Church Calendar to indicate temporal segments in *Erec*, time in *Cligès* is reckoned by months, weeks, days, and hours. . . . secular time systems in *Cligès* suggest a linear conception of human history. Perhaps the best example of such a conception in *Cligès* is the one already implicit in the exordial *translatio* . . Here the concept of time is historical, acknowledging a cyclical rhythm in the rise and fall of civilizations which are displaced in irreversible chronological succession. ("Discourse" 16)

Thus not only is the audience's response to the lovers' elopement and their fantastic *locus amoenus* determined by the work's emphasis on the constraints of mimetic contingency, but at the same time Chrétien defines marriage and adultery themes within the context of wide-ranging social and historical consequences.

It is obvious, therefore, that the relation between history and fiction depends on subtle dynamics not accounted for by conventional definitions of social realism. Indeed, the call for a radical reassessment of the concept of "courtly love" has also come from and been propelled by historical criticism. Studies and claims made by the historian John Benton, for example, have explicitly addressed this problem. Benton found little or no evidence to support the assumption that any significant population of medieval society ever truly tolerated, let alone condoned or advocated, an adulterous ethic or ideology.

First and foremost, adultery remained a grave offense throughout the Middle Ages. The prevailing medieval attitude toward adultery is inscribed in Roman and Germanic law codes, which both provided severe penalties for adultery:

> The penalty provided in the Theodosian Code (XI, 36, 4) was that the vio-
> lators of marriage should be sewed in a leather sack and burned alive, but
> Justinian tempered this to give the woman a scourging and send her to a
> nunnery, from which the husband might release her if he chose (Nov. 134,
> c. 10). The barbarian codes often provide the penalty of death, or permit
> a father or husband to kill an erring daughter or wife, and a husband
> could kill his wife's seducer without incurring vendetta. (24)

And while it appears that medieval laws and attitudes were more lenient
toward the extra-marital forays of men (they being judged as adulterers
only if they seduced a married woman) and that secular courts may have
only infrequently meted out serious punishment, it remains true that "the
tradition of private vengeance was so strong that violators of marriage still
ran grave risks" (26). Benton's sources point to the enduring nature of this
common law approach to the punishment of adulterers throughout the
Middle Ages. Even in the late thirteenth century, says Benton, Beaumanoir
(*Coutumes* XXX, 102) speaks of the husband's right to act as judge and
executioner in the event he should come upon his wife and the culprit *fla-
grante delicto* and act swiftly upon his rage (26). To the extent that
medieval lords indulged their sexual prerogatives, the evidence suggests
they did so most often less in a spirit of sublime sentimentality than in one
of boastful antagonism and one-upmanship (24–5). On the other hand,
lesser vassals who dared to seduce the lady of a liege lord braved the most
terrible forms of retribution imaginable (26–7). Accordingly, says Benton,
there would be little reason for the associates of a lord's court to proclaim
their intimate secrets in public song (27).

Economic considerations could also serve to discourage adulterous conduct. Although displaced by dotal marriage practices, the custom of asso-
ciating the bride with the surrender of a brideprice left its imprint in the
form of the high prestige value that medieval Europeans placed on virginity (Hughes 290). Finally, infractions against the monogamous vow could,
in various ways, adversely affect the assignment of dotal assets (Benton 25
and Hughes 285).

Benton's negative historical critique leads him to reject the idea that
medieval authors and poets celebrated a purely secular form of love
divorced from legal and (Christian) moral precedents, and his views raise
doubts about critical assessments of the troubadour corpus. This is
because, as critics such as W. D. Paden, Jr. and A. R. Press have reminded
us, the troubadour corpus is, itself, inherently deficient in terms of social
realism and conceptual precision:

> In an essential way much of the Provençal *trobar is clus*: the troubadours
> refuse us access to their social and historical setting, permitting us to see
> only the reality they choose to show. (Paden 49)

As Press points out, critical contentions that secular mores condoned adulterous desire or conduct cannont be supported by reference to the purely *incidental* evidence to be gleaned from the lyric corpus (328)—a point which Paden's study drives home with statistical clarity:

> The total number of poems we have surveyed is 503. In 16 per cent of them the poet does not speak of love . . . In another 47 per cent of our texts the poet does speak of love, but there is no evidence that his lady was married or that she was superior to him. (36)

Thus, says Press, critical appreciation of the concepts that constitute *fin'amor* and arguments pertaining to the fundamentally adulterous nature of the ethic are in fact based on *historical* assumptions not literary inferences (331 and 338: "Only when the sociological argument has been put forward or tacitly assumed is the incidental evidence then adduced to support the general theory").

Press shies away from the task undertaken by Benton, namely an attempt to describe in historical terms "the real-life context in which the troubadour love-lyric was cultivated" (Press 331), but his study complements Benton's effort by endeavoring instead to trace the sources of the critical delusion that provided the cornerstone for popular sociological approaches to the understanding of *fin'amor*:

> It is, I think, now clear that if only in one single but absolutely vital aspect, namely the possibility of young noblemen having any direct, personal relations at all with young unmarried noblewomen, the picture drawn by Miss Paget and reproduced by A. Jeanroy and others, unsupported as it is by conclusive, unequivocal documentary evidence, is in itself unreliable and, when contrasted with the full evidence of reliable primary sources, manifestly distorted. But that picture, on analysis, proves to be the sole material basis on which, whether explicitly as in J. Anglade and A. Jeanroy, or implicitly as in J. Frappier and M. Lazar, the sociological argument in support of the theory of adulterous *fin'amors* rests. (Press 338)

Finally, Press points out that erroneous assessments of twelfth-century court life in southern France gained further currency through the kind of pseudo-historical testimonies found in the fictional *Vidas* and *Razos* on the one hand, and Andreas Capellanus's satirical love manual on the other—the former giving rise to adulterous fictions, the latter creating the illusion that *fin'amor* constituted a programmatic ethical code (328–9).

In the final analysis, Press rejects any essential association between twelfth-century love themes and adulterous concepts. Paden argues, likewise, that the main tenets typically associated with troubadour love find no clear support in the corpus itself:

> Two cardinal tenets of received literary history, namely the beliefs that the
> ladies whom the Provençal troubadours loved were the wives of other
> men, and that they were superior in social rank to the poets, rest on very
> few poems . . . We conclude that the currency of these two beliefs reflects
> a tendency of the critic to express his own perspective, not that of the
> poet. (29)

Benton asserts that reference to conceptual formulations of amorous sentiment that depart from the orthodox dichotomy between *caritas* and *cupiditas* reflects anachronistic distortions imposed by modern critics, and he,
like Robertson, rejects the term "courtly love" as a purely fallacious critical construct (Robertson 17; Benton 37.)

Have we come full circle then, back to the tenets of Robertsonian criticism? Actually, Benton explicitly allies himself with Robertson's contention
that medieval authors worked in a fundamentally *ironic* mode when treat-
ing of adulterous love. Like Robertson, Benton focuses on Chrétien's
Lancelot as evidence for this general critical assertion:

> If we find Lancelot a sympathetic figure because he was guided by love
> rather than reason, it is because modern attitudes differ from medieval
> ones in ways Chrétien could not foresee. I am therefore in agreement with
> Professor Robertson that Chétien wrote courteously of Lancelot and left
> him locked in a tower, rather than condemning him explicitly, not because
> he found his behaviour admirable but because he was writing in the
> medieval tradition of irony. (110–111)

For Benton, the love which the troubadours declared was a virtuous sentiment, virtuous in Christian terms, which denoted loyalty, honor and protection but not emotional commitment or carnal desire (109 and 112).
Hence, amorous concepts continue to be circumscribed by the bounds of
reason and Christian virtue, or so it would seem. But if Press finds ample
grounds on which to reject the implicit logic of Lazar's assertions that troubadours advocated *adulterous* love as such, he never goes so far as to question Lazar's equal or more fundamental assertion that troubadour poets
commonly speak of a longing for carnal satisfaction. When Press examines
the relation between lyric and narrative motifs, therefore, it is primarily to
show that the latter in no way provide (even *post facto*) evidence or depictions *from which* to infer the existence of an adulterous ethic. Hence,
Press's incisive logical argument makes no larger claims regarding medieval
attitudes toward love, except to show that love and marriage could be conceived in tandem and that *fin'amors* was not necessarily adulterous. As for
the relation between lyric and narrative traditions, Press's analysis remains
neutral and does not seek to impose any paradigmatic reading of troubadour lyric. Indeed, Benton's own analysis highlights the difficulty involved
in imposing monolithic views on medieval attitudes toward love:

Having reiterated how serious the consequences of adultery could be, I should also stress the toleration of it which was so commonly found in some circles of medieval society. As we have seen, if a husband did not take action himself, adultery often went unpunished and perhaps even uncensored. Medieval literature can be quite complacent about adultery, as in the *fabliaux* . . . or in the song attributed to the Countess of Die, in which a lady sings to her knight that she has "a strong desire to hold you in place of my husband, as long as I have your promise to do all that I would like." (109)[33]

Thus, Benton makes historical and literary observations that complicate his at other times categorical assertions about the Augustinian views underlying all medieval attitudes toward love, as when he contends that the epic refrain made famous in the *Song of Roland*—"Pagans are wrong and Christians are right"—provides an apt model for the dichotomous nature of medieval moral thinking. Surely the famous epic refrain represents a vanishing horizon; the twelfth and thirteenth century are, on the contrary, characterized by astonishing degrees of syncretism and synthesis.

Press suggests what our own observations have shown in greater detail, namely that northern French narrative literature of the twelfth century reveals no inherent tension between the impulses of love on the one hand and the virtues and bonds of marriage on the other (332–3). Paden refrains from making any assertions about medieval attitudes toward love, but asserts that Chrétien's *Lancelot* is—by virtue of its very preoccupation with illicit love—inherently anomalous, and that "Provençal lyric does not provide a background for Chrétien's *Lancelot* in which adultery formed a common theme" (49). Paden's study reorients the search to explain and define secular love themes in the medieval *romans* and returns us to where Benton leaves off when he rejects the concept of a secular, amoral ethos independent of fundamentally theological concepts and dichotomies. Should we return to Augustinian doctrine as the sole basis for appreciating twelfth-century preoccupations with profane love?

Somewhat ironically, this proposition leads back again, in a new way, to the hermeneutic link between literature and social reality by simply asserting a radical revision of contemporary attitudes based not on the "romantic" constructs of modern critics,[34] but upon what is perceived to be the canonical moral doctrine of the Middle Ages.

We must also consider that Chrétien's *Lancelot* is arguably *exceptional* with regard to its flagrant staging of adulterous passion.[35] This text aside, how do we situate other narratives in relation to the devices of didactic irony when they dramatize not adulterous transgression, but rather tales of *marital* love and adventure? Where does moral didacticism intersect with pagan myth and secular themes? Is it fruitful to have recourse to the notion of a Christian moral "gloss" as a means of better understanding the significance of amorous and marital fictions in twelfth-century narrative works?

In an article with the promising title "The Social Significance of Twelfth-Century Romance," R. W. Hanning attempts to obviate patristic dichotomies by attributing to chivalric romance and its public sensibilities and convictions defined in quite modern, secular terms. Hanning argues that "chivalric romance embodies the conviction of its audience that self-consciousness is the key to successful activity in the cause of self-fulfillment, and the awareness of its audience of a tension between experienced private needs and imposed public or external values and obligations" (3–4). Although the anachronistic tenor of Hanning's conceptual terminology is sufficient in and of itself to cause us hesitation, the application of Hanning's alternately dichotomous model—which places an enhanced *value* on private experience—distorts the true nature of the opposition between public and private, external and internal in its attempt to describe amorous plots. We saw this kind of dichotomy operating in *Cligès* and *Partonopeus*, but we saw that it does not reflect a truly antagonistic or exclusive opposition. Indeed, the psychology of love involves a certain embarassment which can lead to a breach between private desire and public conduct, but this breach is elaborated in narrative terms as ironic deferral: the (il)legitimacy of competing unions is revealed and divulged only gradually; in the end, however, only one overarching ideology prevails, and it is this ideology that defines the shifting angles of ironic perspective. Fenice and Cliges become the butt of comic irony only after they choose clandestinity; they do not choose clandestinity because they adhere to anomolous "values." Clandestinity is an expedient and not the outcome of a moral conviction. Nor does any moral "conversion" or abdication precede a harmonious reintegration of the individual to society. Once made public, the union of Fénice and Cligès quickly leads to the indefinite assimilation of the individuals to their established and legitimate roles in public life.[36] In both *Cligès* and *Partonopeus de Blois*, competing models of marital legitimacy are indeed juxtaposed, but the broader opposition between the ideals of subjective desire and the pressures of social contingency is represented not in terms of moral value or subversive autonomy but with the subjective and alienating features of amorous psychology. Likewise the tension between amorous rivals is not truly based on the opposition between marriage and cupidity, but on the opposition between competing models of conjugal legitimacy. In the ascendent model, furthermore, love is in fact a fundamental feature of this legitimacy,[37] such that the impulsive and ostensibly subversive aspects of desire (while rich in ironic and comic possibilities) ultimately drive narrative events toward a culminating restoration or recuperation of legitimate and harmonious order.

Hanning, for example, sees the rivalry surrounding Partonopeus's conjugal destiny as one between public and private values, between external prerogatives and the hero's true nature (21). Yet both marital options arise logically from and lead ultimately to largely identical sets of values and rewards. The mythical logic of the text makes it difficult in fact to separate outer from inner, to the extent that Partonopeus's privileged mythical status as the king's nephew is innately emblematic of his inner nobility. Mélior loves Partonopeus because of his social standing and innate nobility, because of his *honor* (vv. 1353–78; 535ff.). In fact, the archetypal sovereignty rivalry that governs the plot would utterly fail to distinguish between individual nobility and social rank. At the same time, the French poet must have consciously invested the fundamental opposition between two worlds and between two brides with some new and meaningful typological design. We saw that Chrétien created a line of demarcation between mimetic and marvelous discourses as part of his critique of love-sick fantasy. The *Partonopeus* poet delineates two worlds, but in order to delineate an *inner* space of subjective desire, he goes out of his way to minimize any meaningful *external* contrasts: the magic love potion appears not in conjunction with Mélior's marvelous fairy domain but in the context of the court of the Frankish king (Partonopeus's uncle) where it is introduced as part of a crass scheme to lure the young hero away from his mysterious lover and into an arranged marriage which promises immediate physical and economic gratification. While purely tactical motives serve alongside the narcotic effect of the potion to mar the sanctity of the politically arranged (but aborted) betrothal at the king's court, Partonopeus's idealistic liaison in the Otherworld is also marred at the outset by an act of violent rape, and later by Partonopeus's betrayal of his promise to Mélior.[38] At the same time, both realms offer a similar promise of unparalleled happiness. Hanning argues that "the potion stands for the lure of self-forgetfulness, implicit in accepting the good life lived at court" (21, note 45). Of course, this is exactly what Mélior's marvelous kingdom also offers to the extent that it is a place of uninterrupted pleasures isolated from the wars that trouble France (vv. 1331ff.; 1881ff.).[39] The poet also attenuates Mélior's appearance as a supernatural fairy by having the heroine attest to her Christian faith and virtues (vv. 1145–60; 1530–46).

Hence the Otherworld in *Partonopeus* is not a world of alternative, "true" or inverse values, but a world of feudal values and ideals rehabilitated and enhanced by subjective illusions which are inflamed by youthful love and passionate suffering. Indeed, when the bishop of Paris (who conspires in the marital plot along with Partonopeus's mother) gives the young hero the lantern that will illuminate the body of the fairy woman (a sight temporarily forbidden to the young suitor), the metaphorical implication is that the two worlds exist only as a function of an inherent inner illusion, and not according to any true opposition of terms. Hanning somewhat unconsciously stumbles upon the same conclusion when he points out that the amorous crisis eventually leads to a ready resolution of the conflict at which point the lovers discover "a society neither as terrible nor as insurmountable as ignorance and innocence had led them to believe" (22). Thus while the "progress of love" involves the establishment of a metaphorical space of concealed sentiments and inner consciousness, it does not stage a conflict between distinct ethical codes as such, between honor and prowess on the one hand and an inner truth of personal beliefs or aptitudes on the other. The idea that free subjectivity is an absolute value stems only from Hanning's appeal to modern psychological assumptions (13–18). In *Partonopeus*, love is not a value unto itself; it is rather a natural, universal, physiolgical and psycological affect with predictable symptoms and vulnerable to artificial means of stimulation. As desire it provides an inner counterpart (man as microcosm) to the unfolding of historical events. It serves to integrate the actions of the individual / hero to a larger, external scheme of feudal and providential values. We will eventually want to examine how marriage fictions came to play a pivotal role in a dynamic ideological typology in which love and marriage provide the link between the inner world of the hero and the external world of heroic and historic destiny.

In the narratives we have examined so far, love and prowess remain closely linked, and amorous desire (while it may come into conflict with established orders) is in fact integrated into a new vision of aristocratic ideals. Any *perceived* incongruity between internal impulse and external honor represents nothing more than a subjective miscalculation or misperception with regard to competing interests and contingencies— miscalculations and misperceptions rich in dramatic and ironic possibilities. We saw how Thomas transforms the *Tristan* legend into a commentary on the inescapable suffering that ensues from obsessive desire both in and out of wedlock. The doctrinal analogue for this psychological fatalism is overtly dramatized by the "salle aux images" motif in Thomas's *Tristan*. In *Cligès* and *Partonopeus de Blois*, however, we see that the same construct has been amplified and shaded by Ovidian tropes and levity to reflect what must have been perceived as a more rational gloss of amorous motivation, one that could escape outright moral condemnation and ethical catastrophe and resolve itself, more or less painfully, more or less skillfully and judiciously, into a publicly sanctioned and socially beneficial matrimonial bond.

The relation between the archetypal plot elements and authorial perspective still requires further commentary (see chapter four), but for the moment it is interesting to observe that twelfth-century French writers did not necessarily receive the legendary tale of Tristan and Iseut as a paradigm of subversive love. Themes identified in the opening subplot involve a certain psychological realism whose starting point is the typically subjective world of lyric expression, but the text quickly transposes the elaboration of amorous themes to a more pragmatic level of action. The relation here between the lyric and narrative themes is not defined as the working out of a synthesis between antithetical concepts (i. e. love / marriage), nor as a progression from *illicit* love to *marital* love, but more specifically and pointedly as an interest in the proper progression from the initial psychological paralysis of amorous infatuation to a constructive *realization* of amorous desire ("En amors a molt greveuse oevre").[40] We can argue, therefore, that erotic love was not conceived to be an entirely amoral, subjective or artistic construct because it could find its proper place in the hero's itinerary (a potentially positive literary fabulation) and in relation to the customary, sacramental, and institutional requirements of medieval marriage (a cultural construct that involves impulses both social and intellectual, secular and religious). Thus it seems that the itinerary of an amorous hero like Cligès might not be adequately explained by pitting one ethic against another or one concept of love against another. The chivalric knight as both hero and lover had his roots in pagan traditions both classical and Celtic and posed a unique challenge to the medieval poets who sought to carry western literary wisdom and tradition into their own Christian age.

Therefore, even where doctrinal categories remain operative from a moral and conceptual point of view, we must leave open the possibility that marriage fictions serve to dramatize diverse perspectives on sentimentality and ethics and that as fictional *topoi* they are subject to both diachronic and synchronic contextualization. Neither their significance nor function can be determined (either entirely or at all) in a non-linear vacuum devoid of the narrative structures and traditions to which they have been assimilated and, perhaps, subordinated. In twelfth-century thought love was already a syncretic construct incorporating disparate cultural influences, Christian and pagan. Furthermore it had become intimately linked with both epic and marital themes, themes which in the available classical and Celtic models alike played a complex and indispensable role due to the predominance of the sovereignty motif in greater Indo-European tradition. Ovid may provide a new perspective on the Virgilian hero, but the dynamics of love and passion that gave new dimensions to the medieval protagonist would have to find their place within a larger epic and cultural typology whose discursive field could be both that of historical and literary legend. This kind of exegesis is based on the late antique and medieval *figura* and departs significantly from the static typology of the codified exemplum, the magic symbol or abstract cosmological or moral allegory. The *figura* was, as Auerbach points out, "by nature a textual interpretation," which, by its power to transform pagan wisdom into a corresponding form of Christian and providential revelation, was well adapted to didactic aims (as opposed to occult or highly erudite speculation) and which has proved to be the gift of mature, self-assured, and self-conscious cultures (56–7).[41] Although Auerbach identifies Dante's *Divine Comedy* as the culmination of this figurative tradition, it should be pointed out that Dante's extreme synthesis (64: "All these forms [figural, allegoric, and symbolic] occur in the work which concludes and sums up the culture of the Middle Ages . . .")

produced a static world of its own where historical contingency has already been fully transcended.[42] The *figura*, however, finds its full measure of play within the open spaces of historical contingency and discursive unfolding, as is so keenly discerned and eloquently conveyed by Auerbach himself:

> Figural interpretation establishes a connection between two events or persons, the first of which signifies not only itself but also the second, while the second encompasses or fulfills the first. The two poles of the figure are separate in time, but both, being real events or figures, are within time, within the stream of historical life. Only the understanding of the two persons or events is a spiritual act, but this spiritual act deals with concrete events whether past, present, or future, and not with concepts or abstractions; these are quite secondary, since promise and fulfillment are real historical events, which have either happened in the incarnation of the Word, or will happen in the second coming. (53)

The posture of the twelfth-century writer as glossator, therefore, means that we must consider how dynamic cultural exchanges and evolving typologies orient the otherwise archetypal configurations of the aristocratic (secular, ethical and national in scope) marriage plot with its attendant motifs: adultery (the love triangle), elopement, betrothal, consent, rape, quest or conquest, separation and (re-)union.

In approaching these questions, we will begin by examining the importance of the theological dichotomies emphasized by Benton in his historical critique of the concept of "courtly love." With this end in mind, I shall turn to the earliest works in our corpus,[43] works which derive immediately from learned Latin traditions and which form the basis for later narrative tales of secular and aristocratic love. Our immediate purpose will be to see to what extent the thematic range of the medieval translation as gloss is in fact circumscribed (or not) by the conventional moral dichotomies (i. e. sin / virtue, *caritas* / *cupiditas*) of Christian doctrine. Such works, it seems to me, are more likely to be vehicles for moralizing endeavors to the extent that they spring from philosophical traditions germane to the Judeo-Greco-Roman crucible from which Christian doctrines immediately descend. As evidenced by Jean Bodel's famous prologue to his *Chanson des Saisnes*, medieval readers themselves seem conscious of the distinction to be made between the pleasing fantasies of the Arthurian marvelous and the didactic wisdom embodied in the traditions associated with ancient Rome (cf. Fourrier 122):

> Ne sont que trois materes a nul home entendant:
> De France et de Bretaigne et de Romme la grant;
> Ne de ces trois materes n'i a nule samblant:
> Li conte de Bretaigne, cil sont vain et plaisant,
> Et cil de Romme sage et de sens aprendant,
> Cil de France sont voir chascun jour aparant. (vv. 6-11)

[*For a man of learning three traditions count: that of France, of Brittany, and Rome the great. Nor is there any likeness between these three. The stories of Brittany are frivolous and entertaining; those of Rome are filled with wisdom and meaningful edification; those of France are true day in and day out.*]

Our study of two twelfth-century *romans d'antiquité* will provide us with a foundation for exploring further the ways in which our understanding of medieval marriage fictions depend upon these didactic and literary exchanges. Jean Bodel's remarks on Arthurian and classical traditions remind us that twelfth-century writers held the wisdom of classical Roman writers in high esteem.[44]

Another famous prologue, that which precedes a collection of *lais* attributed to a real or fictional personage identified as Marie de France, confirms (by virtue of its own apologetic stance) the lesser status of Breton legend while simultaneously affirming what Bodel also implies, namely that texts from classical antiquity were judged to contain a deeper philosophical and moral significance or *sens*. It is exactly this assumption and its concomitant *practice* (the authorial gloss) that Marie de France seeks to transfer to the popular Celtic legends. When Chrétien writes *Cligès*, he interjects into his Arthurian fantasy references to an archetypal Celtic couple (Tristan and Iseut); he also alludes to another archetypal tale of marital conflict, namely the tale of Paris and the abduction of Helen.[45]

Ultimately, our understanding of *Cligès* and other works in our corpus would remain incomplete without a clear appreciation of how twelfth-century authorial practice both contributed to the popularity of marriage fictions while also potentially defining and transforming medieval perceptions of archetypal characters such as Paris and Helen or Æneas and Dido as they simultaneously laid a foundation for medieval interpretations of the marriage plots fundamental to the larger Trojan heritage. Marital intrigues are central to both the destruction of Troy and the founding of Rome— epic-historical events which retrace the movement of the European *translatio* evoked by Chrétien in the prologue that opens *Cligès*.

NOTES

1. A point which has been illustrated empirically (Teperman 15 and note 5): "Dans la littérature romanesque médiévale, la charpente narrative de bon nombre de récits des XIIe et XIIIe siècles est édifiée autour du mariage et surtout du mariage d'amour. . . . Nous avons constaté que dans le volume 4 du *Grundriss der Romanischen Literaturen des Mittelalters* (1978), qui s'intitule "Le Roman jusqu'à la fin du XIIIe siècle" et qui donne une liste exhaustive de tous les romans en vers de cette époque qui ont été conservés, sur les 58 romans mentionnés, 38 (c'est-à-dire 66%) comportaient le schéma narratif de la thématique de la conciliation de l'amour et du mariage, le reste se répartissant en diverses contes d'aventures guerrières et amoureuses (ces dernières mettant en jeu l'instabilité des amants ou

l'infidélité des époux) sans motif prédominant récurrent. Ce taux vient donc, par là même, mettre l'accent sur le succès que remportèrent les romans sur le mariage d'amour et l'intérêt qu'ils suscitèrent auprès des poètes comme de leur public." Marital plots also take on a growing importance in Old French epics (Teperman 2, 16).

2. Ed. F. X. Newman.

3. On the problem of pinpointing the "height" of marriage doctrine developments and controversy in the twelfth century, see page 11 (and note 7) below.

4. Marcabru, XV, as reprinted in Gaunt ("Marginal Men" 62-3). Gaunt translates: "Restraint comes from speaking in a gracious manner and courtliness from loving, and he who does not wish to be blamed must avoid all unworthy behaviour, deceit and folly, then he will be wise, as long as he thinks about it./ For thus can a wise man reign, and a good lady improve herself, but she who takes two or three lovers, and does not wish to put her faith in one, must indeed make her worth and merit grow always more vile." In the light of the argument that I am about to make, I would prefer to translate *mesura* by "moderation" (i. e. as a positive virtue) rather than as "restraint" (which suggests prohibition and vice).

5. For a similar approach to medieval social classes, see Kim McCone's work on the early Irish *fían* (e. g. *Pagan Past and Christian Present . . .* ch. 9).

6. For an approach to Old French works based on Duby's two models and corresponding literary genres, see (for example) Teperman. It should be pointed out that incipient concepts of marital consent which formed an inherent part of the Church's attempt to affirm the sacramental efficacy of marriage led to unexpected contradictions with regard to both the so-called lay and ecclesiatical models (and this will be the subject of a later chapter). Payen has asked whether or not medieval marriage ideologies would not translate into a neat system of typological correspondances based on literary genre (220), but he ultimately concludes "Il résulte de tous ces exemples que la conjugalité prend des visages très différents selon les catégories sociales, et que le mariage médiéval est une réalité littéraire très diverse dont la complexité n'est réductible à aucun système, qu'il s'agisse de typologie ou d'idéologie" (226). For an example of syncretic possibilities in hagiographic literature see "Amour, mariage et sainteté. . ." in: *Amour, mariage et transgressions au moyen âge*, ed. D. Buschinger and A. Crépin (Göppingen: Kümmerle, 1984) 73-91: ". . . l'hostilité des hagiographes contre le mariage n'est pas absolue . . . L'institution reste suspecte tant qu'on la considère au niveau de l'amour humain et de l'union des corps. Elle prend une autre signification si l'on considère que sa finalité ultime est l'union des esprits dans le vibrant contexte de l'amour divin. C'est à partir de cette démarche dialectique que les hagiographes proposent un idéal matrimonial" (86-7).

7. Duby *Two Models* 19-21; *Dict. de Théologie Cath.* col. 2123-4; Gaunt ("Marginal Men") points out that historians are in some disagreement around this issue: "Duby situates the most intense part of the struggle in the last decades of the eleventh century; Brundage, more convincingly, sees the first half of the twelfth century as the crucial period and locates the final triumph of the church model of marriage during the pontificate of Alexander III (1159-81) . . . From this period

onwards the church model of marriage came to be accepted as the theoretical norm in Christian society and . . . it was recognized that the Church had the right to legislate and adjudicate on matters pertaining to marriage" (55). Brundage has, however, corroborated Duby's perspective to some degree: "By 1100 the Church had secured virtual supremacy in the adjudication of issues relating to the formation of marriage and the separation, divorce and remarriage of those whose marriages failed (Brundage 223). We recall that Gaunt wished to privilege the period 1100-1150 while Shirt wishes to argue that the controversy was most pronounced in the period 1150-1200. At one point Duby gives no more narrow historical time frame for the "principle turning point in the social history of European marriage" than the 10th to 12th centuries (Duby *Love* 13).

8. On William, his poetry and social milieu, see Bezzola 206ff.

9. "Chez Thomas [l'amour courtois] devient un art de souffrir et d'en mourir: c'est dans la mort des amants qu'il atteint son suprême accomplissement . . . Cette vie, la courtoisie l'imprègne de sa vertu, non pas morale, mais esthétique: elle l'embellit des complications les plus exquises, par quoi les âmes d'élite cherchent sans cesse à se dépasser, à s'affirmer, à tendre plus haut vers un idéal de beauté, de charme et de volupté qu'elles savourent . . ." (107).

10. Jonin (291) and Le Gentil ("Tristan" 126) also emphasize the fatalistic aspects of the potion and the inherent opposition between the passion it conjures up and the lovers' ability to act freely and with lucidity. It is surprising that critics have not attempted to elucidate the supernatural motif (i. e. the love potion) in light of its flagrant Ovidian parallel (a parallel we will discuss later).

11. Conforming in this respect to allusions to Mark found in the legend of St. Pol Aurélien (Fourrier 44-5).

12. Professor Joseph Duggan has pointed out to me that in Occitan lyric the husband's role need not be dispassionate or simply hostile. The husband may, like Mark, suffer amorously, as the *gilos*. Yet Thomas's Mark seems to depart from the burlesque figure of the *gilos* that informs Béroul's treatment of Mark and later presents itself in the archetypal portrait of Archimbaut in the thirteenth-century Occitan narrative *Flamenca*.

13. Le Gentil's description of Tristan as an amorous hero who has fallen from grace seems to accord well with Thomas's account of the episode involving Tristan le Nain and his encounter with Tristan l'Amoureux. Tristan le Nain is unable to recognize in Tristan the ideal lover he imagined and idolized. As with Tristan's and Iseut's love, idolatry here turns to disillusionment but with the added irony that it is now the "hero" Tristan who fails to fulfill the expectations associated with his own stature as an heroic lover.

14. Köhler *Idéal*, pp. 163-5 and cf. our discussion below, pp. 25-6. Köhler accounted for the discrepancy in attitudes in economic terms. However, it is difficult to detect in *Cligès* the kind of class dialectic that no doubt interests Marxists but not medieval thinkers and which Köhler evokes as an economic basis for a broader North / South opposition in twelfth-century France: "Dans le Nord, où s'annonce l'alliance antiféodal entre la bourgeoisie et la couronne, la chevalerie se trouve déjà acculée à une attitude défensive . . . "(163). The idea of a "northern" mentality

opposed to a "southern" can provide a facile logic for attempting to reconcile the incongruities between the troubadour corpus and the narrative *romans*. According to Coppin, for example, people living in northern France would be more "realistic" and more religious: "Hommes du Nord, réalistes, positifs, ils entendent pousser l'amour jusqu'à sa conclusion naturelle . . . mais en même temps, la force des idées religieuses et de l'institution sociale leur impose le respect du mariage" (71). Interestingly, the historian J. Benton argues that from the evidence provided by legal documents and records it seems that the communities of southern France actually imposed more severe penalties for adultery than did their northern counterparts (25).

15. This association is made quite explicitly through Fénice's ironically applied and oft-quoted allusion to Paul's text on marriage (*Cligès* vv. 5304-9; I Corinthians VII, 8-9). For a brief discussion and illustration of the assimilation of Ovidian eroticism to Christian perspectives on love in the twelfth century see Baldwin (23-4); more on this perspective in chapter 3.

16. A point made emphatically by Nelson, p. 83.

17. Fourrier describes Cligès's failure to protest against Alys's marriage as an "inconséquence manifeste" in the structure of Chrétien's plot (131). Likewise, in *Partonopeus de Blois*, a physical act actually precedes without hindering an ensuing period of amorous courtship. As for the deferred marriage, it is (Partonopeus's) age and political protocol that are evoked as temporary obstacles. Before making public their intentions and getting married, the lovers must wait until the time agreed upon between Mélior and the nobles of her realm—a date proposed by Mélior herself only in order that Partonopeus might establish himself as a knight (*cevaliers*) before being evaluated as a prospective consort by Mélior's court council:

> Ne m'en tenés à anuiosse
> Se li termes est issi grans;
> Car ensi est li covenans
> De moi à tos cels de m'onor
> Que je doi dont prendre segnor.
> Li termes lor est ennuios;
> Mais je l'ai fait, amis, por vos:
> Por ço que pensoie à vos prendre
> Les ai fait tant longes atendre;
> Car dont ert li termes pleniers
> Que porés estre cevaliers,
> Et dont à primes, à honor,
> Vos porai eslire à segnor.
> Ço ne lor seroit bon ne bel
> Qu'offrise à prendre un tousel . . . [etc.] (vv. 1476ff.)

The problem of (Cligès's) age is also one among other motivations invoked to explain why Cligès and Fénice defer from challenging Alys's right to marriage and the imperial throne; the hero is only thirteen (v. 543) to young to marry, at least according to the norms of medieval canon law.

18. These topics will be examined in depth in chapter three.

19. Or between the lovers' (subjective) perception and external reality. We will explore this dichotomy further in our discussion, below, of *Partonopeus de Blois*.

20. Various mss. differ in their ratio of Occitan to Francien dialectical features, but the most integral and representative mss. (the Oxford ms.—the base for the edition used here, i. e. Combarieu du Grès and Gouiran 1993 which is based in turn on the Mary Hachett edition—and the Paris ms.) both contain preponderantly Occitan traits.

21. Combarieu du Grès and Gouiran p. 7 (citing M. Pfister [325] "La langue de Girart de Roussillon," *Revue de Linguistique Romane* 34 [1970]: 315-25).

22. See Louis, 2: ch. 3 and 3: 73ff. Louis's claims are countered by some critics; for the larger controversy and speculation over the poem's unity and composition see (in addition to Louis) Joseph Bédier, "Recherches sur la formation des chansons de geste." *Les légendes épiques*. 3e éd. Vol. 1. (Paris: Champion, 1926): 3-95; Ferdinand Lot, "Encore la légende de Girart de Roussillon. A propos d'un livre récent," *Romania* 70 (1948): 192-243 and 355-96; and Le Gentil ("*Girart*"). I discuss Louis's theory further in chapter four.

23· The terms "prologue" and "epilogue" used by Louis, and initially applied in Paul Meyer's introduction to his translation of the poem (1884; rpt. 1970, pp. xxxix-xliv), are now in common usage. I adopt them for the sake of convenience.

24. Although, based on Louis's definition of what it is *not*, it is hard to conceive of what he deems *amour vertu* to be. I discuss Louis's view in more detail in chapter four.

25. Heintze has offered a complementary perspective on the structure of *Girart de Roussillon*: "Surtout dans les chansons de geste très tardives, des héros séjournent assez souvent temporairement au royaume d'Arthur ou bien ils y entrent définitivement à la fin de leurs vies . . . C'est pourtant le poète de *Girart de Roussillon* qui longtemps auparavant fait un premier pas dans cette direction . . . " (56-7).

26. The terms of this opposition conform essentially to what Duby has defined as the *lay* and *ecclesiastic* models (of 12th-century marriage practice) respectively. We will see later, however, that the medieval poet of *Girart de Roussillon* problematizes the ideological dichotomy (secular vs. Church) that characterizes Duby's formulation. For a different dichotomy, between predatory and arranged betrothals see Duby's contrast of courtship models within lay society (*Two Models* 107-9). For a study of marriage in medieval narratives with an approach based on Duby's two models, see Teperman.

27. For a thorough yet concise overview of the sovereignty motif in Celtic tradition with reference to its broader Indo-European manifestation, Kim McCone (chapt. 5, esp. pp. 109-116) provides an excellent resource.

28. Susan Crane has taken this approach to insular romances of the twelfth and thirteenth centuries ("The narrative pattern of departure and return that characterizes these works is typically incorporated in a pattern of dispossession and reinstatement . . . By translating a basic revenge pattern into terms of feudal reinstatement and translating love motifs into terms of family stability and continuity, this literature accommodates fundamental Anglo-Norman baronial concerns" (18; for

reference to Celtic features in the narrative patterns, see 18-19; 23; on the exceptionally marginalized status of English barony and for contrast with French nobility, p. 20; and Warren, pp. 232-34; 367-80).

29. There is in fact a distant Greek archetype for the ambiguous status of the sovereignty or cross-over hero in the topos of the noble shepherd who attracts the goddess of love (Paris, Anchises). See Grimal, pp. 13-14. In chapter four I will discuss further the emergence of the Greek Paris as an heroic archetype in the Old French *roman*.

30. In *Girart de Roussillon* and *Tristan* it could be said that the privileged amorous relation actually contributes to the hero's disenfranchisement.

31. For the importance of (pseudo-) historical discourse and historical scope to the logic of typological exegesis and pastoral instruction see Auerbach. We will return to this point below.

32. Duby suggests that in literary representations of courtship and marriage, lay and ecclesiasitical models of marriage prevail according to the genre at hand. The *romans d'antiquité*, however, provide a salient example of the ways in which the medieval *roman* commonly elaborates multiple (pseudo-) discourses that can complicate genre distinctions. Furthermore, despite the neat dichotomy (lay vs. ecclesiastical) underlying Duby's "models," Church reformers seem not to have confronted fundamental lay prerogatives and customs based on family authority and political stability directly. Rather, they attempted to develop a conceptual reorientation of marital legitimacy sanctioned by natural law and effected through the operations of sacramental grace (Colish 2: 637-8, 658). In chapter four, we will examine how writers influenced by emerging doctrine also discredited aristocratic marriage customs and contrasted them with unions based on sacramental principles.

33. To account for these voices, Benton argues that this is a manifestation of another category of medieval mores, one based on the aristocratic and warrior code of honor: "For such people the issue of adultery was not morality but honor, and their major concern was who wore the horns. For every woman who betrayed her husband there was a man who had made a conquest . . ." (27). In this respect, he seems to anticipate Gaunt's ("Femmes") approach to amorous themes in *Girart de Roussillon* based on a *homosocial* theory of troubadour love (cf. pp. 29-30 above). We have already indicated some of the problems with this perspective. Moreover, one has to wonder how this view accords logically with the particular literary examples Benton had alluded to, namely the *fabliau* tradition, the Countess of Die and Aucassin!

34. Benton uses the image of the "backward glance" (36) just as Robertson uses the metaphor of the rearview mirror in his magisterial *A Preface to Chaucer* (Princeton: Princeton U P, 1962) and cf. Robertson "Courtly Love" 17.

35. We pointed out earlier that Celtic traditions already provided for tales of marriage, adultery and abduction, and traditional material would again seem, as in the *Tristan* legend, to provide for a tale of adultery. Among the Welsh Triads, for example, is one referring to Guinevere's betrayal of Arthur: "Three Faithless Wives of the Isle of Britain . . . and one was more faithless than those three: Gwenhwyfar, wife of Arthur, since she shamed a better man than any of them." (Bromwich 1961;

Triad 80). What is interesting in Chrétien's *romans* is how he exploited the popular Ovidian tropes of his day to refashion the pagan tale in a way that generates humorous ironies but which also erects a new medieval archetype that assimilates the Ovidian *passio* to the Christian passion theme.

36. As Noble points out with regard to Fénice and Cligès, it is the very mental habits of honor, pride, loyalty, and deference that in fact lead the lovers themselves to defer from acting on their impulses (40).

We will see that the portrait of Briseda in the *Roman de Troie* (see chapter two) may shed some light on Chrétien's depiction of Fénice. In any case, I will discuss Fénice's role and Chrétiens' ironies in more detail in chapter four.

37. In the twelfth century marital affection was, of course, deemed an integral feature of marital legitimacy (more on this in chapter three).

38. For the mother's attempt to lure Partonopeus along with the help of her uncle into an arranged marriage see vv. 3919-4050. For the end of the sequence (which recounts Partonopeus' betrayal of Mélior) see vv. 4343-4556. For Partonopeus' initial meeting with Mélior and the "rape" scene, see vv. 1243-1330. In chapter four, I will discuss these scenes in greater detail.

39. Paralleled by only slightly less grandiose promises of wealth and happiness in the context of the Frankish court (vv. 4009-24).

40. According to the testimony of the Norwegian *Tristramus Saga*, it is exactly this kind of dynamic that provides the foundation for the cautionary aspect of Thomas's *Tristan* as well: "Quel besoin avons-nous d'en dire plus à ce sujet, puisque tous ceux qui ont quelque discernement doivent savoir que la coutume des amants est que chacun d'eux cherche à concrétiser ses désirs amoureux au plus vite, même s'ils doivent se rencontrer en secret?" trans. Daniel Lacroix, *Tristan et Iseut: Les poèmes français, la saga norroise* (1989), p. 519.

41. The twelfth century, in fact, witnessed an intense interest in and a dramatic proliferation of new and finely shaded exegetical viewpoints. For an overview, see Beryl Smalley, *The Study of the Bible in the Middle Ages* (Oxford, 1952).

42. Which is simply to say that Dante's pilgrim encounters historical figures little different from portraits in a gallery; they belong to and are defined (figuratively) by a historical or literary memory / context but they now have a fully consummated (i. e. static) figurative identity and are abstracted from the process of historical and discursive action and unfolding. By contrast, there is Chrétien's prologue to *Cligès* in which, as Maddox has pointed out, the references to the western *translatio* cast an open-ended and potentially ironic contingency on the historical triumph and destiny of France: "Secular time systems in *Cligès* suggest a linear conception of human history. Perhaps the best example of such a conception in *Cligès* is the one already implicit in the exordial *translatio* . . . (vv. 28-33). Here the concept of time is historical, acknowledging a cyclical rhythm in the rise and fall of civilizations which are displaced in irreversible chronological succession" (Maddox "Discourse" 16). Maddox contrasts time in *Cligès* with the cyclical nature of time marked by reference to holy days in *Erec*. The distinction is weakened, however, by the way in which linear time threatens to conform to a similar paradigm, hence the potential irony with regard to nationalistic views of historical destiny. In any case, it is

important to point out that Maddox's allusion to *Erec* suggests that marriage plots were certainly not always part and parcel of a (pseudo-) historical narrative and typology. At the same time, however, it is interesting not only that *Cligès* contains an important historical dimension, but also that its characters and events are associated, by intertextual allusions, to a broad range of historical heroes whose fates are bound up with those of national history.

43. The next chapter section will begin an analysis of two *romans d'antiquité*.

44. This could be true even among the theologically minded when confronted with a master of profane love such as Ovid; see Baldwin's discussion, pp. 20-22.

45. A passing reference is also made to "la guerre Polinicés" (v. 2519) providing another allusion to Greek legend through the *Roman de Thèbes*.

CHAPTER TWO

Love, Marriage, and Mythology:
Marriage Fictions and Epic History

The *romans d'antiquité* are extended narratives in French verse dated to the twelfth century. Among the earliest examples are the *Roman de Thèbes*, the *Enéas* and the *Roman de Troie* which have been dated to the 1150's and 1160's.[1] Each is a narrative composed of octosyllabic couplets adapting material handed down from classical pagan antiquity. Scholars have also suggested that the *Thèbes*, *Enéas* and *Troie*, in addition to being closely linked in time, may share political themes influenced by the historical marriage of Henry II Plantagenet to Eleanor of Aquitaine in 1152 and by the subsequent expansion and dynastic claims of the Anglo-Norman court. The *Enéas* is now considered to have preceded the *Roman de Troie*, and the *Roman de Thèbes* seems to be the latest of the three works.[2] Among the *romans d'antiquité*, our study of marriage fictions will focus on the *Enéas* and the *Roman de Troie*.

The *romans d'antiquité* initiate an expansion of twelfth-century literary horizons by consciously supplementing the historical, rhetorical and human dimensions of feudal epic with literary conventions (deriving principally but not exclusively from Ovid) that provide the foundation for a fuller vision of human motivation and historical agency which inevitably comes to define a new cultural ethos. In his study of the *Enéas*, Lee Patterson asserts that its medieval author only minimally exploits the wealth of literary possibilites created by the innovative application of disparate literary conventions to a revision of historical epic. We shall see that these same possibilities are thoroughly if not exhaustively probed by the author of the *Roman de Troie*. The most urgent question governing our discussion, however, will be *how*—from what perspective, within what larger vision—the authors of these two ostensibly epic works envisaged the integration of amorous and sentimental motifs.

In the case of the *Enéas* and the *Roman de Troie*, each author was working from a source whose plot involved important episodes relating to mar-

riage and illicit love: Æneas, Dido and Lavinia on the one hand, Paris and
Helen, Agamemnon and Clytemnestra on the other.[3] To what extent do the
medieval versions go beyond purely anachronistic and rhetorical amplifi-
cations of these amorous subtexts to arrive at a coherent and dynamic gloss
that allows for the preservation of epic themes in tandem with the exploitation of new modes of lyrical discourse? Do the marriage tales merely super-
impose a Christian moral teaching based on Augustinian dichotomies onto
pagan legends? How effectively did the medieval authors negotiate the stylistic and thematic synthesis of such disparate genres? To what extent does
our appreciation of amorous intrigue in the context of these appropriated
epics provide a cogent foundation for understanding the interest in love
and marriage in the chivalric *romans*?

Hanning, for example, sees a profound discontinuity between the
romans d'antiquité and the chivalric *romans* which he based on the dishar-
monious clash of lyric and epic concerns. He asserts that the motif of the
amorous knight has only a marginal and superficial presence in the *Enéas*
where "claims for the autonomy of chivalric heroism inspired by love could
hardly be made in such a context of large and impersonal forces" and
where the relation of the chivalric topos to the "overplot" remains purely
"disruptive" (8–9). Likewise, some critics have noted the somewhat unnatural and truncated presence of the Ovidian themes and rhetoric as they are
adapted by the author of the *Enéas* (Auerbach 215; Yunck 48; 259–60).

Yet medieval authors do not seem to recognize such rigid distinctions
with regard to literary theme and genre. For example, a similar but inverse
dialectical tension reveals itself in what Hanning wants to categorize as
quintessential "chivalric" narratives. In *Partonopeus de Blois*, for example,
the poet has grafted epic, historiographical, and dynastic dimensions on to
his amorous hero's marvelous itinerary. In *Cligès*, likewise, Chrétien consciously integrates pseudo-historical dimensions (Maddox 1982).
Eventually, we will have to ask ourselves how these apparent paradoxes
can be resolved if we are truly to appreciate the significance of the marital
and amorous intrigues elaborated (upon) by medieval authors.

Most criticism of the *romans d'antiquité* has indeed focused primarily
on sources or on the technique of translation itself, narrowly defined.
These studies foreground dominant stylistic features, but they tend to
ignore the larger thematic designs that motivate the creative impulses
behind the activities of *clerc* and "translator." Even what is esteemed as
"new," namely the lyric matter, typically goes back to Ovid. At first glance,
the *romans d'antiquité* appear void of meaningful innovation, mere *pas-
tiches* of disparate literary models. Faral's seminal study goes a long way in
showing the degree to which the amorous episodes that constitute "original" additions in the *Enéas* and the *Roman de Troie* are constructed like
modern buildings with pre-fabricated blocks of conventional Ovidian rhetoric (*Recherches* 133–6). But, while Faral takes note of the formative liter-

ary influence exercised by the *romans d'antiquité*, he simultaneously expresses little appreciation for the artistic merit of these same works which he himself credits with making France the center for an enduring literary *translatio*.[4]

Faral did acknowledge, however, the inevitably *atomizing* effect that source studies like his have, and he looked forward to the kind of monographs and thematic studies that have appeared during the past four decades.[5] Focusing predominantly on lyrical episodes and amorous motifs, however, most of these thematic commentaries tend to create a fragmented image of the larger narrative structure of the *Enéas* and *Troie*. This is to be expected, however, given the novelty of the lyric elaboration that provides these epic stories with their most innovative features. Marriage, a not uncommon theme in Old French epic, is also a prominent theme in the *Enéas* and *Troie*, but the introduction of a new amorous discourse inspired by Ovid seems to alter its significance.[6] The *chansons de geste* provided the closest literary analogue for the retelling of pagan epics, but in borrowing from troubadour poetry and the lyrical Ovid, the authors of the *Enéas* and *Troie* imported discursive registers often indifferent or antagonistic to the various dynastic and economic interests associated with medieval aristocratic marriage strategies. Likewise, the language and postures of lyrical love, by virtue of their metaphorical, abstract and potentially duplicitous nature, are essentially at odds with the literal and binding nature of speech and action in the world of the *chansons de geste*.

No doubt the apparently facile symmetry of the *Enéas* (which builds upon the twin panels of Eneas' unhappy and happy amorous encounters and exploits the neatly dichotomous fortunes of the two Latin heroines Dido and Lavinia), like that of the *Roman de Troie* (whose cycles of internecine strife are perpetuated by marital transgressions and epitomized in the forceful abduction of Helen by the Trojan knight Paris), suggests and invites a ready-made "sens" condemning infidelity and celebrating marital propriety. Behind the apparently reactionary treatment of classical visions of *eros*, we naturally assume the presence of the moralist, and the logic of the moralist is of course bound to the dichotomous relation between love as cupidity or sin and love as charity. Inherent in this perspective, which focuses on desire and subjective choices, is a belief in the proper and meaningful relation between individual (mis)conduct and individual (mis)fortune. The intimate coexistence of love and marriage fictions in classical legends no doubt prompts moral judgments on the part of twelfth-century writers with clerical education, and numerous critics see the *Enéas* and *Roman de Troie* as expressing a moral-didactic gloss on love and desire. Most studies focus on ethical and moral questions oriented around the proper progress of amorous desire in the direction of spiritual refinement, monogamy and enduring marital fidelity.[7]

These assumptions lead, in the case of the *Enéas*, to a nearly unanimous consensus concerning the fundamentally bipartite structure of the poem's

narrative form and moral oppositions: "Retardant à l'envi la réalisation de
ce mariage riche de promesses historiques, le roman médiéval nous présente
d'abord son envers illicite: la liaison d'Eneas avec Didon."[8] If indeed the
idea of a purely secular amorous ethos is largely a construct of modern-day
criticism, and critics such as Benton and Robertson are right to define
twelfth-century attitudes toward love in fundamentally Augustinian terms,
it seems natural to see the classical epics and the amorous or marital transgressions of their main protagonists as ready-made *exempla* providing
material for an edifying Christian typology of sin and virtue. Do the marriage fictions in the *Enéas* and *Roman de Troie* conform to such didactic
purposes? How do these twelfth-century appropriations of classical legend
and chronicle establish parameters for our appreciation of the Trojan subtexts—centering on the meaning of Paris's seminal "judgment" and memorialized marital transgression—to be evoked in later works such as
Chrétien's *Cligès* or Gautier d'Arras's *Eracle*, or Jean Renart's *Guillaume
de Dole*?

ENÉAS

> C'est en cela que consistait l'*immoralité* d'Ovide, et non dans la vivacité
> ou l'indécence de ses peintures. Il révélait à son siècle ce que celui-ci avait,
> déjà, confusément aperçu, qu'il n'y a pas un amour 'permis' et des amours
> 'tolérées,' mais, ainsi que l'avait, après Lucrèce, écrit Virgile, que l'amour
> est le 'même pour tout ce qui vit,' que la passion a ses racines dans l'être
> même et n'est pas une maladie ou une aberration honteuse.
>
>   P. Grimal, *L'amour à Rome* (164–5)
>
> [*This is what accounts for the* immorality *of Ovid, and not the explicit
> nature or indecency of his descriptions. He revealed to his century what it
> had, already, dimly perceived—namely, that there are no "permitted" and
> no "tolerated" forms of love; rather (as Virgil himself, after Lucretius, had
> written), love is the "same for all living creatures," and passion has its
> roots in man's very being and is not a disease or a shameful aberration.*]

And that my miseries may increase the more, the greatest part do not so
much respect the value of things as the event of fortune, and they esteem
only that to be providentially done which the happy success commends.
By which means it cometh to pass that the first loss which miserable men
have is their estimation and the good opinion which was had of them.
What rumors go now among the people, what dissonant and diverse opinions! . . . The last burden of adversity is that when they which are in mis-
ery are accused of any crime, they are thought to deserve whatsoever they
suffer.

                    Consolation of Philosophy I. pr. iv. 154–163.[9]

For Virgil, marriage seemed to provide a unique occasion for imagining the possible reconciliation of instinct with reason, of individual desire with public order.[10] Ultimately marriage can be seen as the pivotal institution that serves to assimilate the individual to the stabilizing continuum of (aristocratic) genealogical extension and to the course of historic destiny itself:

> L'ensemble des textes qui concernent [le mariage] occupe, dans les recueils juridiques, une place considérable, qui nous garantit l'importance attachée à un acte don't on attendait d'abord la survie mais aussi, et peut-être plus encore, la stabilité de l'Etat . . . . (Grimal 63–4)

> [*The totality of texts treating the topic of marriage account for a considerable part of the judicial collections, proving the importance attached to an act that was relied upon, not only for the survival, but perhaps even more, for the stability of the State.*]

At the end of the *Aeneid*, however, Virgil provides only a tenuous glimpse of an uncertain future destiny. The eventual marriage of Æneas and Lavinia is never celebrated, and Virgil's narrative ends abruptly with the slaying of Turnus. This vengeful act, motivated by private rage, casts a shadow on Æneas's "progress" and re-inscribes the founding of Rome within the cycles of irrational violence and destruction that serve as hallmarks of the tragic vision informing the Greek world (Patterson 163–5; Spence 43). Thus, Virgil's text casts a shadow on the very imperial promise ratified in the mythic dimensions of its fictional vision. As the *Enéas* poet sought to adapt Virgil's work to a contemporary, secular audience, he seemed to have recognized the subtle but appealing and immensely significant aspects of the marriage theme. In what is perhaps its most original emendation of Virgil's text, the twelfth-century *Enéas* provides a description of Æneas's wedding day. In so doing, it seems to offer a more optimistic ending in which the images of war yield to the pomp and splendor of a joyous imperial marriage and cycles of destruction give way to the satisfying promise of a prosperous and peaceful line of dynastic succession:[11]

> Quant vint al terme qui mis fu,
> c'a grant poine orent atandu,
> li rois ot ses amis semons
> et mandé ot toz ses barons.
> Contre Eneas est fors issuz;
> a grant joie fu receüz,
> a Laurente l'an a mené
> voiant als toz l'a erité
> de son realme, de s'enor;
> tot li a otroié lo jor
> que il sa fille a esposee.
> Grant leece I ot demenee . . .
> Eneas fu a roi levez,
> a grant joie fu coronez,

et fu coronee Lavine:
rois fu d'Itaire et el raïne. . . .
Eneas ot le mialz d'Itaire,
une cité comence a faire . . .
Longues l'a Eneas tenue,
puis est an sa main revertue
tote la terre al roi Latin . . .
Ascanïus regna aprés,
et puis fu si com Anchisés
a Eneas ot aconté
an enfer, et bien demostré
les rois qui aprés vendroient . . .
Molt furent tuit de grant pooir
et descendirent d'oir an oir . . .(10091–150).

[*When the deadline arrived (which had been eagerly awaited), the king
summoned his friends and all of his barons. He sallied forth to meet Eneas
and was received with great joy. They led him to Laurentium surveying as
they went the wealth and holdings of his realms. The king granted to him
everything on the day he married his daughter. The day was filled with
celebration . . . . Eneas was declared king. He was crowned amidst great
joy at the same time Lavine was crowned: he became king of Italy and she
queen . . . . Eneas had the best lands of Italy and began to build a city
. . . . Eneas held the reign for a long time, then all the land reverted to the
possession of the king of Latium . . . . Later Ascanius reigned, and it came
to pass just as Anchises had foretold it to Eneas in the underworld when
he clearly named the kings who would follow . . . . All of them were very
powerful, each one succeeding the previous heir.*]

Where Virgil questions the Roman myth of cultural progress by inti-
mating the resurgence of an irrational and tragic mythology within the
workings of a historical *logos*, the medieval author contrasts the happy
nuptial rites of Eneas and Lavine with the tragic consequences of the Dido
/ Didon episode in a manner that suggests a more optimistic view of historical destiny governed by a new divine covenant.[12]

An index to the imaginative significance of marriage in history and to
the poet's own conscious investment in the marriage theme comes at the
very beginning of the *Enéas* when the poet reminds us that the providential
destiny unleashed by the downfall of Troy and culminating in the marriage
of Eneas and Lavine was itself set in motion by an earlier marriage drama:

Quant Menelaus ot Troie asise,
onc n'en torna tresqu'il l'ot pris;
gasta la terre et tot lo regne
por la vanjance de sa fenne.
La cité prist par traïson,
tot cravanta, tors et donjon, . . .
tote a la vile cravantee
a feu, a flame l'a livree.

Li Greu prenent les citeains
nus n'eschapot d'entre lor mains . . .
Ocis I fu li reis Prianz
o sa fame, o ses anfanz:
unc ne fu mais tant grant ocise.
Menelaus a vanjance prise:
toz fist les murs aplanoier
por le tor fait de sa moillier. (vv. 1–6; 9–12; 19–24)

[When Menelaus assaulted Troy, he did not relent until it was taken. He laid waste the land and all the kingdom to exact revenge for his wife. *He took the city by treason. He destroyed everything, towers, fortifications . . . . The whole city he destroyed, letting it be consumed by fire. The Greek soldiers took the citizens; none of them escaped their hands . . . . King Priam was killed along with his wife and children. Never again would one witness such terrible slaughter.* Menelaus got his revenge: he razed all the walls of Troy on account of the wrong suffered by his wife.]

In this passage, the poet does not misrepresent the underlying motive of the war, but he recasts the motivating incident in exceptionally domestic terms and portrays the destruction of Troy as a private war of vengeance presided over by Menelaus himself.[13] He also omits any mention of intermediate atrocities, so that the offense done to Menelaus is held before the reader's eyes as the sole cause and continuing impetus of war from the beginning of the Trojan conflict up until the final collapse of the city walls. The passage gains additional weight by its *marked* position at the opening of the narrative. The lack of further reference to the Paris / Helen story and the author's own silence on the moral issues surrounding the abduction of Helen also engage the listener's own judgment while leaving the exact moral implications of the event unresolved. Finally, the exclusive import given to Menelaus's leadership serves as a counterpoint to the work's final marriage scene. In other words, the narrative is framed by two pivotal marriage dramas, an archetypal tale of abduction and revenge on the one hand, and a regal and idyllic conjugal union on the other.

The emphasis on private love and private marital conflict has led numerous critics to approach the *Enéas* as a moral gloss of Virgil's epic. The contrast between the "sinful" overtones of the Didon episode and the properly sanctioned nuptials of the Lavine episode seems to provide a ready structure for the dramatic representation of the hero's journey from near perdition to redemption, from sinful *amour-passion* to the willful cultivation of a spiritually inclined *fin'amor.*[14] The desire to read the *Enéas* as a "morality play" bears directly upon any final conclusions we will draw concerning the motivations behind and potential reception of the motifs of marriage and adultery.

Approaches based on a rigid moral dichotomy between *cupidinous* and "spiritual" or *charitable* love, have larger narratological implications as well. An important feature of the critical reading which sees in the *Enéas* a "moralized" progress of love is, as we have just suggested, the underlying assumption that the narrative is fundamentally divided into a bipartite structure which manifests itself in a number of symmetrical dichotomies between a) Didon's "passion" and Lavine's "love"; b) between the Trojan abductor (Paris), whose crime initiates a cataclysmic war, and the Trojan exile (Eneas), who lays the foundation for a new empire; c) between the Eneas who encounters Didon and the Eneas who ascends from the land of the shades to go on to a happy marriage:

> The Dido episode, a superlatively evil and idolatrous experience, *narcis-sistic,* will give way, in the second half of the romance, to outward love. . . . The selfishness, solitude, and negation of this first part of the Enéas will contrast profoundly with an overload of *selflessness,* positive sacrifice, and solidarity in the second half. (Cormier 141)

The central distinction between the Didon and Lavine episodes depends, for Cormier, upon the opposition between "solitary" and "communal" love—the former being marked by improper forms of passion (*"sapientia-*turned-folly"), the latter by a virtuous and reciprocal love that "defeats erotic anarchy" and "beats alienation" (138). In the end, according to Cormier, the love between Didon and Eneas is a "dead end" (141, note 39), while the love between Lavine and Eneas allows for moral consent and wisdom (Cormier, 253). Marriage, accordingly, proceeds from and serves to celebrate and reward wise and spiritually enlightened love. On the level of narrative, the happy marriage dramatizes the redemption of the past as it is carried out through a process of moral enlightenment through time.

Ostensibly, one should need only look to the respective fortunes of the two heroines as proof of the larger moral lesson they serve to exemplify. It is indeed difficult not to see the *Enéas* as a kind of textbook study of cupidinous vs. enlightened love, and more recent arguments advanced by Barbara Nolan make a powerful case for such a reading. Whereas Cormier (131–141) focused on the term "amor soltaine" (appearing in the epitaph on Didon's tomb, ll. 2139–44), Nolan scrutinizes the term "foolish love" in the same epitaph ("ele ama trop folement") (159). For Nolan this is con-vincing evidence that Didon is to be understood in doctrinal terms as a fig-ure of "illicit" love as it is explicitly defined in twelfth-century glosses of Ovid in opposition to "chaste" (married) love. According to Nolan, school commentaries on Ovid provided "a neatly formulated theory of foolish love and legitimate marriage. And it was this theory . . . that gave shape and *sen* not only to this poem but also, *mutatis mutandis,* to most other classicizing romances of the later Middle Ages" (159). Nolan adduces a number of compelling arguments in support of her views. She points out that Didon herself is made to echo the scholastic formulations when she calls herself *fole.* This is taken as a sign that the poet is evoking the scholastic topos of *amor stultus* which the glossators evidently derived from the adjective "stulta" employed by Ovid in the *Heroides*. This scholastic topos teaches that a woman is foolish to love a man who will go away (165). For her knowledge of the scholastic glosses in question, Nolan relies (as I do) on Hexter's erudite study (1986). In his study, Hexter remarks that at least one glossator draws a connection between the opposing forms of amorous conduct and their respective consequences: foolish love brings on misfortune, while benefits ensue from legitimate love. Nolan echoes this point (170), and her argument naturally leads her to infuse her reading of the *Enéas* with the same moral bias and the same "exemplary" logic:

> In a parallel way, the *Enéas*-poet uses the stories of Paris and Dido to exemplify the nature as well as the unfortunate consequences of illegitimate and foolish love. Then, in the last long segment of his poem, he introduces a counter-example. If Paris and Dido exemplify illicit and "fole" love driven by exclusive, unbalanced, private desire, the *Enéas*-poet uses the last movement of his poem to provide a third, corrective example in his close, celebratory study of Lavine's courtship and marriage. Instead of preferring love alone, as Paris and Dido have done, Lavine and Eneas enter into a social, politically suitable love. (170)

In some sense both hero and narrative are cleaved in two. The Eneas who is figured by the sinful actions of his kinsman Paris manifests himself in the events leading up to and involving the Didon episode. The Eneas who serves as a positive example of Trojan superiority emerges in the second half of the narrative.

This exemplary logic, however, ultimately conflicts with the poem's larger narrative strategies. It is clear that the *Enéas*-poet expects us to lament Didon's fate, but the attempt to read the narrative progress in terms of a simple and dualistic moral division only leads to multiple contradictions and distortions. First, we must reconcile the sinful Didon with the observation that the *Enéas* poet retains something of the original aspect of Virgil's "optima Dido" (Bk. IV)—a contradiction that Cormier has observed but not resolved.[15]

Secondly, we also need to explain how this implicit moral opposition and the narrative discontinuity that it necessarily implies are to be reconciled with the logic of the work's genealogical myth and historical typology—a logic that assumes a transcendent *continuity* through generations and events. This continuity is indispensable to the legitimizing functions of the genealogical fiction that links Troy to the twelfth-century Norman-Angevin empire. Finally, it is prefigured in the Trojan identity common to both Paris and Eneas (which inevitably envelopes the entire Trojan story in layers of potential irony).[16] This continuity is highlighted by the Platonic vision of metempsychosis adumbrated in Virgil's Book VI, and is (curiously) preserved, in its essential outlines, by the *Enéas* poet's own description of Anchises's speech to Æneas. This plane of linear connections transcends the ostensible dichotomy between hero and anti-hero and ostensibly provides the basis for Patterson's claim that the *Enéas* poet has purposefully set out to "renegotiate" the historical vision by expunging the darker shadows of the past and presenting in the positive outcome to the Latin wars a myth of cultural progress and imperial stability (180–181).

Yet Patterson's perspective does not *resolve* the dichotomies, it *erases* them by attributing such an erasure to the poet's own strategic aims.[17] At the same time, however, the internal logic of the narrative, namely the emphasis placed on genealogical continuity, seems to problematize Patterson's exceptionally optimistic characterization of the work. For if the author of the *Enéas* sought to erase the dark side of Virgil's hero, why would he explicitly underscore this historical continuity at the very moment one would expect him to downplay it—at the culminating point of the narrative, during the description of Eneas's triumphant political conquest and wedding ceremony?

> Contre Eneas est fors issuz;
> a grant joie fu receuz,
> a Laurente l'an a mené,
> voiant als toz l'a erité
> de son realme, de s'enor;
> tot li a otroié lo jor
> que il sa fille a esposee. . . .
> Eneas fu a roi levez,
> a grant joie fu coronez,
> et fu coronee Lavine:
> rois fu d'Itaire et el raïne.
> *Unques Paris n'ot graignor joie,*
> *quant Eloine tint dedanz Troie,*
> *qu'Enéas ot, quant tint s'amie*
> *en Laurente;* ne quida mie
> c'onques deüst avoir nus hom
> an tot lo mont tant de son bon. (10095–10114)

> [*He sallied forth to meet Eneas and was received with great joy. They led him to Laurentium surveying as they went the wealth and holdings of his realms. The king granted him everything on the day he married his daughter. The day was filled with celebration . . . . Eneas was declared king. He was crowned amidst great joy at the same time Lavine was crowned: he became king of Italy and she queen.* Even Paris never knew a greater joy when he held Helen inside the walls of Troy than Eneas now had, holding his beloved in Laurentium; *it would be hard to imagine that any man in all the world could ever have so many of his wishes fulfilled.*]

This provocative reminiscence and comparison complicates the "moralist" perspective. How do we reconcile the similar passion of the two famous lovers with the disparate fortunes they experience? Why would the two famous Trojan heroes, alike in their Trojan descent and Trojan attributes, not represent a common destiny and common election?

As for twelfth-century scholastic glosses of Ovid, Hexter expresses a similar query in relation to the implicit assumptions made by Nolan who asserts that glossators traditionally identified the *utilitas* of Ovid's epistles (i. e. to warn students against illicit love) in terms of the misfortunes suffered by illegitimate lovers:

> One can hardly take the commentator at his word here. Of the women involved in "legitimate" love affairs enumerated above only Penelope (and perhaps Hermione) enjoyed some benefit from their love following the writing of the letter. . . In fact, in very few cases is there any mention of the subsequent fate of the letter-writer or addressee, so that in practice Ovid's ironies were neither perceived nor flattened. (Hexter 158)[18]

Any perceptive reader (medieval or modern) of Ovid can hardly fail to notice the charm of these ironies. Phyllis makes the painful discrepancy between the lover's merit and the lover's (good or bad) fortune a virtual leitmotif of her letter (*Heroides* II). Phyllis's "Dic mihi, quid feci, nisi non sapienter amavi? / crimine te potui demuervisse meo" (vv. 27–8) is rich in ironies which can clearly cut both ways, and Ovid plays on the ironic relations between private conduct and private fortune when Phyllis makes reference to her own exemplary conduct toward Demophoon as a premise for anticipating a happy outcome and for having believed in her husband's fidelity (e.g. 45–7, 105–116). Her altruism is, however, ironically subverted by her amorous motives. But Ovid adds an essential pathos to Phyllis's plight by having her allude to the formal marriage vows which ratified her relationship with Demophoon ("fides ubi nunc . . ." vv. 31–44).[19] In the end, Ovid's ironies are not moralistic but cultural; they make us sympathize with Phyllis because of her naiveté and because of the rustic passions and the rustic mores of her own land (Thrace) which make her into the victim of a savvy seducer from Athens. In similar fashion, Ovid's Dido vainly imagines Æneas being duly punished by the gods for his crime of deceit (*Heroides* VII; vv. 64–72).

The difficulties of glossing a stoic universe in terms of schoolmarm morality are also aggravated by the conventional, static, and secular nature of Ovidian tropes, taxonomies, and physiology.[20] As lovers or better yet "love sufferers," Didon and Lavine are more alike than different. It is, after all, Cupid who (following Venus's command) fosters Didon's love, and Lavine's passion is described in the same conventional terms used to describe Didon's suffering.[21] Didon "swoons and sighs," "shivers and shakes." "She suffers much and passes the night in great sickness and pain" (vv.1219–57). She is simultaneously "healthy" and "sick" (vv. 1270–75).

Lavine's trials are identical:

> Ele comance a tressüer,
> a refroidir et a tranbler,
> sovant se pasme et tressalt,
> sanglot, fremist, li cuers li falt [etc...] (vv. 8073–6)
>
> [*She begins to perspire, to experience chills and to tremble again and again, to swoon and to faint. She sobs, shudders; her heart fails her, etc.*]

Furthermore, Lavine is (like Didon) alienated from her own will by a compulsive passion incited by Cupid (vv. 8622–59). Nolan, once again evoking glosses of Ovid, argues that Didon is judged to be "foolish," since she loved without reciprocation (165). But Lavine also ventures to love without proof of Eneas's feeling and in spite of the hero's unpromising status as a kinsman of the notorious Paris and as a reputed sodomite (vv. 8567–8611)![22] Moreover, Eneas's departure is portrayed not as a selfish action but as a pious one: although his men are eager to leave Carthage, it would have pleased Eneas to stay: "molt li plaüst li remanoir, / mais il s'an va par estovoir, / si cum li dé l'unt esgardé" (vv. 1655–57). Didon *accuses* him of betrayal and indifference, but Eneas expresses a sincere love as he describes his dilemma (vv. 1757–84).[23] In the passages alluded to by Nolan, therefore—when Didon describes herself as a "foolish lover" (e.g. v. 1814, v. 1850, v. 2047)—we are forced to recognize a self-incrimination made in a moment of outrage and based on moral accusations (i. e. "my love went unrequited") that misrepresent Eneas's larger predicament. Lavine's love is also not without its own madness—she, again like Didon, threatens to commit suicide should her love go unrequited.[24]

Finally, the author of the *Enéas* chooses to moderate the lascivious and bestial overtones that animate Virgil's depiction of Dido and Æneas (*Aeneid*, IV. 129–174). We no longer hear of the "primal Earth" or the "foaming bit" (*Aeneid* IV. 135 and 166). And although for a medieval audience the hunting motif was, in itself, a familiar literary trope with an array of exemplary possibilities, the *Enéas* poet does not exploit Virgil's account of the lovers' excursion to cast a moral judgment on Didon.[25]

Largely alike in their amorous pathos, the two heroines also share quite similar moral and marital dilemmas. Lavine betrays her betrothed lover when she falls in love with a foreign prince in the same way that Didon's love for Eneas leads her to betray her faith to her deceased husband. While critics reproach Didon for betraying her faith to Sychaeus, the poet only mentions this point in passing and goes to some length to justify Didon's marital aspirations. Most notably, Anna encourages her widowed sister to remarry and take the Trojan prince as her new husband, because (she reasons) he will help her to defend her land (vv.1326–1368). It was by no means uncustomary for widows of landowners to remarry, and the custom was echoed in a number of twelfth-century fictions (e.g. *Raoul de Cambrai*,

Thèbes, and *Yvain*) (Jones 39).[26] The medieval poet also attenuates Didon's apparent moral offense by emphasizing her past fidelity: "Onc mes puis la mort son seinor/ ne fist la dame nul hontage" (vv. 1528–9). The point is forcefully reiterated and lamented later by Didon herself (vv. 1991–1998). On the day of Didon's funeral, everyone (including, it seems, the barons and rival suitors whom she offended) come to praise her "courage, wisdom and wealth" (vv. 2125–28). Christopher Baswell attempts to account for the apparent shift in attitudes toward Didon by arguing that "once [Didon] is dead, and her capacity for shifting and transforming value is thus permanently controlled, she regains the praise of her people . . ." (198). This novel interpretation ignores the larger dimensions of tragic irony and historical determinism at work in the poem's broader historical scope. Adumbrated in the plight of the ill-fated individual, however, Didon's fate inevitably takes on overt, but misleading, moral significance. This temptation to orient Didon's plight in terms of overt moral admonishment is tenacious and it continues, in subtle ways, to govern Baswell's own commentary despite his own acknowledged intent to demonstrate the importance of the epic and imperial themes governing the narrative and superseding the amorous motifs.

Consistent with this perspective is the assumption that the major figures are depicted as active agents and movers of history whose destinies reflect moral postures and choices, who directly merit their various deserts. Baswell, for example, obliquely identifies Eneas grammatically but without any real psychological and moral content as the central mover of the narrative and historical events (emphasis added):

> *Enéas's* approach to empire, then, moves from Dido, to battle, to Lavine. *He* passes out of the realm of a threatening feminine "angin" and dominance, and a form of dilation which proves to be solitary, mad, and suicidal. *He* moves thence through the exclusively male preserve of the poem's military center, particularly as exemplified by Nis and Eurialus. And *he* returns finally to a link to woman, "angin" and dilation, but now hierarchically inscribed within the processes of patriarchal dominion, lineage, and logic. (218, emphasis added)

Obviously, there is no justification for supposing that Eneas acts upon a specific doctrine ("approach") of manifest destiny. It is simply that Eneas's itinerary evidently serves, at least on one significant level, to dramatize and perhaps *exemplify* broad tenets of emerging Norman imperial domination and order along with its concomitant threats.[27]

But Baswell's grammatical construct underlies the inherent temptation we are seeking to explore, that a medieval rendering of the Virgilian epic will be governed necessarily (one might say *subconsciously*) by an exegetical habit of mind so inseparable from moral allegory that even the themes of patriarchy and aristocracy (so well appreciated by Patterson and Baswell) will themselves ultimately find their meaning according to the dynamics of personal moral agency and moral exemplarity. But perhaps we should say *dramatize* rather than *exemplify* when speaking of Eneas' fortune. Does not Didon's fate, however glossed, remain predetermined within the context of providential history that underlies the narrative progress as a whole? With regard to Eneas, what subjective dynamic would account for *his* progress from a flawed to a superior moral position and ideology? In his treatment of Virgil's Book VI, the *Enéas* poet gives no resonance to the moralizing fabulations offered to him by Latin commentaries. Ultimately, it is patriarchal lineage that seems to separate the damned from the saved. The poet's description of *enfers* resembles at times a fire and brimstone sermon on the torments of Hell, but it is hard to make sense of the moral warnings with regard to Eneas and Didon. There are giants who sinned through excessive pride, seeking to assert their authority over that of the gods and one who sought to sleep with and dishonor the goddess Diana (vv. 2733–8). The passage through hell, one would expect, would be exploited to dramatize Christian moral dualities, to adumbrate a penitential dynamic *vis-à-vis* Eneas, but it is not. Again, as in the lyric passages, the emphasis is on passive suffering. Eneas is relieved from trials and torments, not discharged of any sin. Piety seems to precede and sanction his passage (vv. 2839–52). Likewise, virtue is also defined in only the most abstract of terms (vv. 2809–10). The real moral conundrum comes, however, in the conflation of virtue and genealogical affiliation. *Caritas* appears as an operative trope: Eneas is tormented by his father's image and leaves his fleet and crew to descend into the underworld (vv. 2857–62). What he learns there, however, is that the Elysian Fields are peopled by Trojans, both past and present. Thus the implicit typological link (i. e. father / Father) upon which the *caritas* trope depends comes to be elaborated not in terms of moral prescription, but in temporal, dynastic and secular terms as a myth of racial pre-eminence (vv. 2811–27).

Thus a prescriptive and pedestrian moral typology looms over all of Eneas' experiences, but the larger vision that constructs the threat posed by Didon depends in fact upon a myth of providential origin and providential destiny that privileges *historic* determinations over *private* judgments and misfortunes. Eneas does not redeem his Trojan ancestry, he fulfills its eternally ordained promise through passive toil and through submission. In the end, the rigid dichotomy between *cupiditas* and *caritas* gives way to a more dynamic mythological typology in which desire drives heroic action simultaneously on two levels, the spiritual *and* the temporal, in which private desire is in no way transformed or suppressed, but its possible channels and objects circumscribed by a higher order, a political (more than merely public) and providential destiny, a supra-private vision.[28]

Didon's misfortune cannot be understood in terms of moral judgment alone, therefore, no more than Eneas can be said to represent a private morality superior to that of his ancestor Paris, both of whom find their place in an elected line of ethnic descent. Likewise, the fall of Troy cannot be, according to the transcendent logic of history as myth and narrative, imputed to a moral transgression and misfortune since it initiates the founding of Rome, and by extrapolation, the Norman conquest. Here, the commentary tradition might provide us a clue as to the nature of the overlaps and distinctions at hand. Although the term "double truth" wasn't coined until the time of the university controversies of the thirteenth century, twelfth-century thinkers inherited a similar notion from Macrobius, and Averroës (1126–1198) made clear distinctions between theological and philosophical articulations of truth just as neo-platonic teachings distinguished between temporal fortune and providential design. At the beginning of his commentary on the *Aeneid*, Bernardus Silvestris refers to the *gemine doctrine* of Macrobius who distinguishes between poetic fiction (*ficmentum poeticum*) and the truth of philosophy (*veritatem philosophie*). Accordingly, any student of the *Aeneid* must not fail to take into account its double teaching. Indeed, those works which are essentially conceived for the purposes of moral instruction are defined as the works of satirists ("Poetarum quidam scribunt causa utilitatis ut satirici"); works for entertainment (*delectatio*) as comedies. This exemplary utility is only a superficial aspect of Virgil's work, however, its overt layer of teaching. As a "history" it not only involves elements of instruction and entertainment (*utilitas* and *delectatio*), it also contains hidden within it a deeper vision which makes it a philosophical text, exploring philosophical truths that transcend static prescriptions and lead toward self-knowledge (*Bernardus* 3 ll. 9–19). Nancy Partner has defined a similar distinction in terms of the narratological paradigms governing twelfth-century historiography—a genre to which the twelfth-century *Enéas* was readily assimilated. Partner points out that the untimely death always provided a ready context for a sermon on vanity with which to "edify the conventional moralists" (218; cf. 58), but that narratological structures, to the extent that they govern the presentation of history, rely on broader, more fundamental rhetorical strategies and philosophical themes. Indeed, while a historical event might solicit a moral admonishment, the larger structures of cause and event usually were based on a deferral—away from private judgment and understanding to a transcendent sphere of order and intervention. This underlying vision of history involves, above all, a doubling of perspectives: "the *saeculum* was theoretically dignified by an invisible cloak of the divine, and the most ordinary happenings might express, in some fragmentary way, an intention beyond immediate appearances . . . ." (Partner 188). Although a commonplace notion, this view of history contained the germ of a posture inherently at odds with conventional moralizing judgments. Speaking of William of Newburgh (who wrote during the second half of the twelfth century), Partner says:

The tradition of historical writing in which William worked demanded also that an historian attempt to be a higher critic and judge who "reads" morally as he records; William's thoughtful reserve before that responsibility gave his work balance and reflective depth. His respect for the profundity and obliquity of the divine mind made him unwilling to blot out the complexity of human experience for the sake of an edifying (and simplifying) lesson. (Partner 52)

In the end, might not it be accurate to say that Didon is as much a character who suffers a *tragic* misfortune, more than exemplifying a *moral* vice? For according to her epitaph, Didon (the good pagan) lacks not *morally* but *culturally*: her predicament stems, not from excessive lust but from an inadequate *savoir*:

> Un epitaife I ont escrit;
> la letre dit que: "Iluec gist
> Dido qui por amor s'ocist;
> onques ne fu meillor paiene,
> s'ele n'eüst amor soltaine,
> mais ele ama trop folemant,
> savoir ne li valut noiant. (vv. 2138–2144)

> [*They wrote an epitaph on her tombstone; the text reads: "Here lies Dido who took her life for love. There would never have been a better pagan if only she had not had a secret love. But her love was foolish; alas, her wisdom was of no avail."*]

It might be possible that the evident "accusation" expressed in her epitaph is as misguided as her own earlier accusations leveled against Eneas, that it represents not so much a final judgment of her character (i. e. "final" from the perspective of the critical reader) as it does the human impulse to fit the inscrutable operations of fate within a "logical" scheme of moral order (i. e. logical within the sphere of immediate experience) in which individual destiny is isolated from a broader horizon of historical contingency just as the moralizing exemplum in the Middle Ages typically takes the form of a decontextualized citation. Like other imaginary epitaphs in medieval literature, might not Didon's epitaph reduce the complexities of historical experience to a "neat" but overly simple and static gloss?[29]

Perhaps as a first step to better understanding the providential significance of the marital dramas in the *Enéas*, we should turn back to Virgil's own account of Dido's crime. Virgil, far from condemning Dido's morality, questions the political and social acumen of her action in the context of aristocratic norms of marriage. Her misfortune is not so much associated with any norms of private moral conduct, but seems to result from a transgression of religious dimensions that is directly linked to the sanctity of marriage itself:

ille dies primus leti primusque malorum
causa fuit. neque enim specie famave movetur
nec iam furtivum Dido meditatur amorem;
coniugium vocat; hoc praetexit nomine culpam. (IV. 169–172)

[*That day was the first day of death, that first the cause of woe. For
no more is Dido swayed by fair show or fair fame, no more does she
dream of a secret love: she calls it marriage and with that name veils
her sin!*][30]

Although in a much more frivolous register, Ovid provided his own readers with admonitions of an essentially similar stripe as he urged them to
remain mindful of the crucial distinction between playful dalliance and
crimes of adultery committed against prominent families and aristocratic
matrons. Dido's "sin" (*culpa*) or "secret love" is one matter; her attempt to
call it by another name (*coniugium*) and thereby confer upon it both a
grave civic and religious sanction (only amplified by her exceptional rank)
is a far graver transgression. Grimal comments that erotic desire and dalliance were typically viewed with indifference by Roman citizens, except in
those cases where such sentiments threatened the prerogatives of aristocratic marriage (e.g. 87, 105–7, 176–80). Furthermore, the sanctity of
marriage was not only tied to the concern over the purity of blood and the
integrity of the *gentes* and its genealogical continuity; it was—even in its
most canonical representation—an institution that integrated private life to
supra-personal orders, both civic and divine:

Contraint de définir le mariage, les juristes ont imaginé une belle formule,
qui situe l'institution à sa vraie place. Nous la trouvons dans le *Digeste,*
où il en est fait honneur à Modestinus, qui vivait au début du IIIe siècle de
notre ère. "Le mariage . . . est la mise en commun du droit divin et
humain," et, un peu plus loin, il ajoute: "Le mariage est l'union totale de
toute la vie." (Grimal 66; cf. 263–4)

[*Forced to provide a definition of marriage, the legal scholars came
up with a convenient formula, one that gives the institution its rightful
place. We find this formula in* The Digest, *where recognition is given to
Modestinus who lived at the beginning of the third century C.E.
"Marriage is the integration of divine law and human law," he says,
adding a little later on: "Marriage is the total union of all one's life."*]

In Virgil's text it is clear—despite the *moral* outrage of the public over
Dido's broken vows—that Dido's unhappy fate is ultimately tied to
the most far-reaching implication of the proposed "marriage": its consequences for Æneas' own preordained role in history. The fatalism that
pre-empts any meaningful moralist reading of Dido's predicament is underscored by the extensive intervention of divine wills.[31]

This clash of perspectives, between moral and providential causality, is even inscribed within Virgil's own text; nor will the medieval poet fail to follow Virgil's lead. In the *Aeneid*, the outraged King Iarbas, spurred on by wild Rumor (*Fama*), alerts Jupiter to the cause for Æneas's "distraction." The *Omnipotens*, however, ultimately shows little sympathy for Iarbas' sense of bruised ethics. Jupiter is solely concerned with divine Providence and Æneas's failure to heed his own preordained destiny (IV. vv. 198–237). The *political* ramifications of Dido's conduct, on the other hand, are immediately highlighted by Virgil's digression on the flight of *Rumor* (IV. vv. 173–197). Rumor's intervention determines the meaning of the event by immediately transforming it from a personal act to a public one which incites jealousy and bitter incriminations. In the end, however, the sinister and volatile nature of Rumor itself raises questions about the value and reliability of "moral opinion."[32] It is not clear, for example, whether the disapprobation conveyed by the reference to "long winter nights spent in wanton ease and without regard for the status of the realm" is an accurate judgment to be attributed to the poet or an evocation of the aspersions of Rumor and the *vox populi* (IV. 193–195). The equivocal nature of the medieval text, with its integumentary layers and complex convergences of fiction and truth, of moral fact and philosophical truth is anticipated in the ambivalent nature of Rumor's revelations ("tam ficti pravique tenax quam nuntia veri" v. 188). The medieval adapter seems to have appreciated the narrative ambivalence: in a subtle but telling fashion he alters Virgil's text to express a subtle commentary on the distortions of Christian moralists. Where Virgil describes the *rapidity* of winged Rumor, the poet of the *Enéas* diverges slightly from his source to emphasize Rumor's capacity for *exaggeration*. Where Virgil's creature clings equally to lies and truths alike, the Old French counterpart ultimately exaggerates so much that truth is transformed into pure fantasy, falsehood and lies.[33]

This subversion of the moral incriminations surrounding Dido's and Æneas' passion is important because it presents love as a *natural* and *universal* impulse and shifts our critical gaze away from gradations of moral propriety and toward a *cultural* dynamic central to the *translatio* of dominion from Greece to Rome.[34] It shifts our interest first to the proper *forms* that regulate the tensions between private passion and public dominion and second to a sphere of providential agency whose designs transcend rigid moral prescription and customary norms. Neither the *Enéas* nor the *Roman de Troie* elaborates a quintessential doctrinal morality regarding sinful and spiritual love. Instead, it seems that the *romans d'antiquité* integrate the themes of love and marriage to a vision of secular history. We will see that marriage fictions play a key role in the integration of secular history to a non-temporal order of divine will.

Before pursuing these questions let us turn from our analysis of Didon and Eneas to see how Benoit de St.-Maure's representation of marital transgression provides for an even bolder inversion of conventional moral expectations. Obviously, Helen and Paris are the most famous and problematic couple in Benoit's narrative of the Trojan war, and we will want to examine how Benoit sought to "gloss" this archetypal tale of abduction by exploiting an innovative synthesis of epic and lyric motifs. However, the Medea story which introduces the Latin Troy narrative provides for more direct parallels with the Didon episode; both episodes relate a tale of a hero who encounters, unites with, and then abandons an assertive and powerful bride figure dwelling in a foreign land. Let us examine the Medea episode in Benoit's *Roman de Troie* before continuing our study of marriage in the *Enéas*.

BENOIT'S JASON AND MEDEA (*Roman de Troie*, vv. 1715–2044)

By all indications, the original text for Benoit de Sainte-Maure's *Roman de Troie* corresponds to the Latin translation of two Greek "chronicles" of the Trojan war (the *Ephemeris Belli Troiani* and the *De Excidio Troiae Historia*).[35] The first of the two chronicles, purportedly written by a certain Dictys of Crete, dates to the fourth century AD but appears to be based on an earlier text in Greek going back to some time between 66 AD and 250 AD and extant only in fragments. Another chronicle of the war, coming down to us only in a medieval Latin version dated to the early sixth century AD, is attributed to a Greek author Dares and also appears to be based on an earlier account in Greek. Both works distance themselves from Homeric conventions and provide a foundation for an "anti-Homeric" tradition which allows for an on-going cultural appropriation of Homer's epic material. Indeed both authors claim to have been eyewitnesses to the events and credit their military "chronicles" (similar in form to Ceasar's account of his military campaign in Gaul) with an accuracy superior to that of Homer's account. The historical pretensions of the two accounts are reflected in their respective titles. Rejecting supernatural features of Homer's epic, both chronicles reorient the Troy legend according to the rational perspectives of Classical historiography. Furthermore, their critical perspective on Homeric heroism provides a model for a medieval tradition which comes to fruition in Benoit's *Roman de Troie* where the conflict leading to the Fall of Troy takes on a pro-Trojan bias and is appropriated for the purpose of contemporary dynastic myths evoking distant Trojan ancestry. The dialectic tension and emblematic significance of the cultural tension between Greek and Trojan which will become a prominent motif in the work of Benoit can also be traced back to the complementary and polemic relationship between Dictys and Dares. The first witness aligns himself with a Greek hero from Crete in Agamemnon's army, while the second claims to have lived within the walls of Troy and served as a retainer to Antenor. Finally, by presenting Homer's great hero Achilles in a new and critical perspective, Dictys, and to a greater extent Dares, provide a model for what we have begun to see as the emergence of novel treatments of the Trojan hero (traditional anti-hero) Paris in specific twelfth-century narratives. Thus while Benoit's affinity for Dares's version should come as no surprise, it should be pointed out that the chronicles also complement one another. In Dares, Benoit found a truncated account of the legend of Jason and Medea incorporated with the events of the Trojan War, but he must rely on Dictys in order to complete his history with his own report of the Greek homecoming.[36] Taken together, therefore, the chronicles comprise a wide-angle view of the war, beginning with a truncated account of Jason's quest for the Golden Fleece that serves as a pretext for hostilities between an earlier generation of Greeks and Trojans which then escalates, after the abduction of Helen, into the full-scale siege of Troy and gives way to tales of exile and return. Benoit's thirty-thousand-verse narrative spans the same broad chronology. It also includes an expanded version of the Jason-Medea episode and numerous lyrical interpolations.

Despite the thin thread of causation which links the tale of Jason and Medea to the rest of the narrative, Benoit amplifies the laconic references in his sources through the use of fairy imagery and Ovidian discourse in order to tie the episode thematically and emblematically to the larger historical legend. Indeed, Dares's limited interest in the episode leaves Benoit relatively unconstrained.[37] If, as it is generally assumed, Benoit had Ovid's telling of the Medea story at his disposal (*Metamorphoses* VII, vv. 1ff.),[38] his telling lacks no less in originality for it. For this reason, and in light of the episode's marked position at the outset of the narrative, we can expect Benoit's creative elaboration of the legend to provide a revealing overture to the work's larger thematic orientation.

To see Medea in terms of a moral gloss on the Ovidian trope of the brazen damsel would be easy. It is Medea who "takes the first step"(1314–1324), and at first glance she appears (as Didon does in the *Enéas*) to be a "foolish" lover who succumbs to a mad passion and brings about her own undoing.[39] At one point, the narrator's description of Medea encourages such a reading:

> Grant folie fist Medea:
> Trop ot le vassal aamé.
> Por lui laissa son parenté,
> Son pere e sa mere e sa gent. (vv. 2030–33)
>
> [*Medea acted recklessly; she had loved the vassal too much. She left her own family for him—her father, mother, and her kin.*]

As the passage proceeds, Benoit clearly departs from the Ovidian model when he shifts his focus away from Medea on to Jason. Now the author voices a less stereotypical condemnation and one made doubly "authorita-

tive" by an oblique attribution to Dares which is later seconded by an interjection in the narrator's voice:[40]

> Assez l'en prist puis malement;
> Quar, si com li Autors reconte,
> Puis la laissa, si fist grante honte.
> Et l'aveit guardé de morir:
> Ja puis ne la deüst guerpir.
> Trop l'engeigna, ço peise mei;
> Laidement li menti sa fei.
> Trestuit li deu s'en corrocierent,
> Que mout asprement l'en vengierent. (2034–42)

> [*He deceived her quite badly. For, just as the author tells us, he then aban-
> doned her, causing her great shame. And she had saved his life; never
> should he have left her. It pains me that he deceived her so. Despicably, he
> made false promises to her. All the gods were angered by this, and they
> would avenge it bitterly.*]

While Medea, like Didon, seems to suffer for acting too quickly and boldly as a lover, Benoit actually encourages us to sympathize with her.
Compared with the nefarious intentions of Jason's avaricious uncle,
Medea's regard for the young hero betrays a sincere and generous love.
Next, Benoit presents a Medea who is clever but innocent at heart.[41] Like
optima Dido, Medea is also noble and virtuous ("sage e gente e bele"). The
marvelous bed of silver and gold (1551–1571) reinforces the other "fairy"
aspects of her character, and when Benoit emphasizes that the young
Medea is fully "worthy" to sleep there, it suggests not only a noble character but a prestigious destiny. In other words, the Celtic features that
inform her exceptional status, reorient alternative traditions for our reception of Medea.[42] The fairy portrait easily blends with the black magic and
exoticism characterizing the classical Medea, but allows Benoit to shape his
heroine into a benevolent figure as he ultimately suppresses the tale of her
magical exploits used to avenge Jason's betrayal. At the same time, the
associations with sovereignty intrinsic to the Celtic bride figure and bride
quest motif make Medea into a figure of commanding rank worthy of
Jason's respect and devotion. She is not *simply* a naïve maiden given over
to infatuation and futile delusions, and Benoit makes it clear that she represents a mate worthy of Jason's own rank. Moreover, the Celtic topos
gives a mythological and providential dimension to her marital union with
Jason. We will soon see how this aspect merges with Benoit's larger historical fabulation; first, however, we should examine how Benoit actually portrays the marital covenant uniting Jason and Medea.

Benoit's attitude toward his protagonists as well as his overriding interest in the marriage theme come to the fore in his original account of the
lovers' exchange of marital vows. Ovid reports on Medea's intent to extract
a firm pledge of fidelity from Jason, but her intentions are overshadowed

with fatalistic doom from the outset. Benoit downplays the certainty of any tragic outcome, and presents the marital intrigue with fresh psychological detail, such that the reader is led to appreciate more intimately Medea's savvy as well as the nature of the deception she suffers. Medea requests and receives not only a lover's pledge, but also a betrothal promise:

> Dame . . .
> Sor toz les deus vos jurereie
> E sor trestote nostre lei
> Amor tenir e porter fei;
> A femme vos esposerai,
> Sor tote rien vos amerai.
> Ma dame sereiz e m'amie,
> De mei avreiz la segnorie:
> Tant entendrai a vos servir
> Que tot ferai vostre plaisir.
> Menrai vos en en ma contree,
> Ou vos sereiz mout honoree. (vv. 1429–40)

> [*Lady . . . . In the name of all the gods and all of our laws, I swear to love you and keep my word. I will take you as my wife and love you above all else. You, my lady, will be my beloved and will be my lord. My intent will be to please you in all things. I will take you to my country, where you will receive great honor.*]

Immediately, however, we begin to realize that the promise is too good to be true. Blinded by love, Medea fails or refuses to recognize Jason's ulterior motives after she spontaneously but imprudently offers her magic powers along with her love. Jason's duplicity soon becomes obvious. His litany of promises has only too perfectly satisfied, item for item, the essential conditions of courtship enumerated earlier by Medea.[43]

But Jason's use of duplicity grows into a more serious offense and Medea's confidence gains in credibility as the lovers proceed. At the courteous, nocturnal rendez-vous with Medea, Jason speaks first and offers, without any prompting on Medea's part, what would seem to be an equally spontaneous, uncoerced and unequivocal vow:

> Dame, li vostre chevaliers,
> Icil qui quites senz partie
> Sera toz les jorz de sa vie,
> Vos prie e requiert doucement
> Quel receveiz si ligement,
> Qu'a nul jor mais chose ne face
> Que vos griet ne que vos desplace. (vv. 1602–08)

> [*Lady, your knight, who is now yours, unconditionally, for the rest of his life, graciously requests that you freely bestow upon him such honor that henceforth he never do you any harm or displeasure.*]

At this point a more noble and prudent Medea reveals herself beneath the impressionable and naïve young girl. As if to test Jason's sincerity, she now refrains from mentioning her magic powers and downplays the importance of any mutual "exchange" or benefit (beyond sincere affection) that might entice Jason to pledge falsely. In addition, (and despite the formal vow Jason has just pronounced), she asks for yet more assurances along with a visible token or *proof* which will ratify the pledge:

> Grant chose m'avez mout pramis:
> Se vos le voliëz tenir,
> Ne me porriëz plus offrir.
> Seürté vueil que jo en aie:
> Puis atendrai vostre manaie. (vv. 1610–14)

> [*You have promised me a great deal. If only you will keep your word, then there is nothing more that you could offer me. It is only some guarantee of your word that I wish for; then I will wait for you to fulfill your pledge.*]

Jason says he wishes to do everything possible, and with absolute honesty, to assure her that she has no cause to doubt him (vv. 1615–18). This time Medea's serious intent is evident, and Jason, likewise, goes well beyond the shallow promises that often find their way into desultory love talk. The vows are then formally consecrated in a solemn ceremony presided over by an image of the almighty Jupiter:

> "Jason," fait el, "venez avant.
> Vez ci l'image al deu des cieus:
> Jo ne vueil mie faire a gieus
> De mei e de vos l'assemblee;
> Par ço vueil estre aseüree.
> Sor l'image ta main metras,
> E sor l'image jureras
> A mei fei porter e tenir
> E a mei a prendre senz guerpir;
> Leial seignor, leial amant
> Me seies mais d'ore en avant." (vv. 1624–34)[44]

> [*"Jason," she said, "come forward. See here this image of the heavenly gods? I have no desire to play games anymore with regard to our mutual union. With this rite I want to be made sure. You will place your hand on the image, and on it you will swear to always act in good faith towards me and to take me as your bride forever. Henceforth, be my loyal lord, my loyal lover."*]

Thus Benoit goes to great lengths to underscore explicitly the fact that the lovers' clandestine coitus is preceded by a formal wedding agreement based on consent, a promise of future affection, and a visible symbol of love and marriage. The culminating sexual act, therefore, is not only legitimate, it serves as the final act which fully consecrates the marital union

(Autre celee ne vos faz:/ Se il en Jason ne pecha,/ Cele nuit la despucela; 1646–48).[45] The equivocal language, however (Se il en Jason ne pecha), once again suggests the gap between Jason's conduct and his hidden intentions. The immediate and obvious cause of misfortune, however, is identified quite explicitly:

> Jason ensi li otreia,
> Mais envers li s'en parjura
> Covenant ne lei ne li tint:
> Por ço espeir, l'en mesavint. (vv. 1635–8)
>
> [*Thus Jason granted her that, but what he said was perjury. Neither vows nor contract did he honor. Bad fortune came to him as a result.*]

This overt condemnation of Jason provides for a revealing comparison and contrast with the otherwise similar structure of the Eneas-Didon story.[46] First of all, Medea's status as unwise lover is clearly, in terms of its moral implications, attenuated by the crime imputed to Jason. Medea, like Didon, suffers from *amor soltaine*, but Benoit does not in any way emphasize Medea's fate in terms of a moral incrimination that then adds to her misfortunes.[47] In addition to incriminating Jason in moral terms, however, he highlights the parallel with Eneas's own adventure by placing Jason's own bride quest within the context of a broader historical destiny. Unlike Eneas (whose fate exonerates him for his abandonment of Didon), however, Jason's actions are not attenuated by the historical perspective. On the contrary, Jason seems to defy the gods and the divine will by betraying a just marriage bond. This is already hinted at in the final narrative remark above (*Por ço espeir, l'en mesavint*), but is made more explicit (as we have seen) in the reference to the anger of the gods (*Trestuit li deu s'en corrocierent, / Qui mout asprement l'en vengierent*). Thus not only does Benoit shift the blame to Jason, he eclipses the monstrous aspects of the classical Medea by ascribing acts of future revenge to the anger of the gods themselves. In so doing, he also distances Medea's private misfortune from any exemplary function (i. e. as a consequence of moral impropriety) and orients it according to a larger, more tragic fatalism based not on amorous deviance, but on an implicit recognition of the intimate link between marriage and historical destiny.[48]

Therefore, in reading the *romans d'antiquité* through the lens of amorous mores essentially oriented in terms of Augustinian moral dichotomies, we inevitably encounter difficulty in our attempts to reconcile the adaptation of "psychological" or Ovidian tropes into an epic framework of heroic action and historic destiny. In his analysis of the *Enéas*, for example, Hanning has pointed out that no mention is made of Lavine at all during the final encounter with Turnus and during the duel itself. Rather, the poet retains the epic motive: Æneas's desire to avenge Pallas's death (8, note 22). For Hanning this is further evidence that the medieval author is unable to unite his hero's *chivalric* motivation (i. e. as an amorous rival) to

his epic exploits (8). We will soon see, however, that this critique is based
on inaccurate assumptions about the conception of erotic love central to
the *romans d'antiquité*. Hanning once again illustrates how the Ovidian
tropes have become inseparable in the minds of modern critics with competing categories of amorous discourse if not with entirely anachronistic
psychological perspectives (Hanning 8: "The *Roman d'Enéas* converts its
epic hero's victory into an exemplum of chivalry heightened in significance
by being linked to the idea of self-discovery through love"). This is patently in contradiction with the emphasis on passive suffering that is a hallmark
of medieval Ovidian love (cf. Baldwin). Perhaps the *romans d'antiquité*
represent a step toward a new literary vision of medieval knighthood based
on the individual quest and a chivalric ethic, but to read them *retrospectively* is to misunderstand their unique and particular position in an evolving, and (I would argue) quite dynamic progression of intellectual responses to the search for new typologies capable of integrating a challenging
array of pre-Christian wisdom marked not only by the affective vocabu-
laries of Ovid, but by the mythological dimensions of both classical and
Celtic legend.

HISTORY, HISTORIOGRAPHY, AND ESTHETIC UNITY:
The Role of Eros

> "hic amor, haec patria est" (Aeneid, IV, 347)

> "Et caelo imperitans amor" (Consolatio II. metr. viii.)

Despite the innovations that reveal a learned posture and elaborate a conscious authorial gloss of the Latin material, we see that the twelfth-century author does not necessarily respond to the ostensible and egregious
moral transgressions of pagan heroes with a predictably conservative moral
critique and condemnation. On the contrary, the fascination or preoccupation with love seems to reflect a progressive intellectual impulse that seeks
to identify and elucidate a more transcendent level of wisdom capable of
spanning the gap between pagan text and history on the one hand and
Christian thinking on the other.[49]

From our analysis so far, there has begun to emerge a vision of love and
marriage defined in terms of a synthetic typology, in which the importance
of epic marriages—marital intrigues around which will pivot the destinies
of entire nations—seems to motivate the medieval author to *uncover*—discover beneath the poetic integument—the proper place of amorous and
conjugal activity within a broad scheme of (now revealed) providential wisdom. Logically, and as our initial analysis suggests, this intellectual posture
would involve a dynamic and creative process of syncretism in order that
poetic fictions based on old orders of values and elaborated within alien
contexts might be assimilated to ostensibly "New-Testament" truths. We

saw evidence of this process in the authors' attempts to reconcile the (potentially tainted) agents of historical destiny with the providential features of their monumental achievements. To some extent the same process manifests itself in the authors' handling of the heroic portrait: the hero(ine) is not morally redeemed through a process of internal revelation and moral edification; (s)he is inseparable from the external and well-attested actions that broadly circumscribe his or her potential value (i. e., as a *type* conveying a no less fixed if latent wisdom) and to which new value is ascribed through various forms of poetic innovation.[50]

We have also noted, however, that a strong desire remains among critics of the *romans d'antiquité* to view the trials and fortunes of amorous adventure in terms of moral exemplarity. Daniel Poirion has warned against overly reductive "moral" interpretations of the *Enéas* (1976 215: "le discours moral . . . [semble] orienter le lecteur vers une édification plus spécifiquement politique"), but he too, while he proves slightly less evasive than Baswell with regard to the emblematic importance of Venus, has difficulty placing the Didon and Lavine episodes on a single, cogent plane of political-historical typology. According to Poirion, Eneas's descent to Hades "liquide l'épisode de Didon par l'expérience . . . purificatrice de la mort" (1976 226–7). R. J. Cormier employs even more evocative language: Eneas, he argues, is transformed by something akin to the touch of "grace" and is infused with an ethical "will to act" (187).[51] However, Eneas had already (previous to his descent) expressed pain and repentance when forced to abandon Dido (vv. 1629–40), and while he learns something about his destiny from Anchises, he seems not to develop any new concept of love itself. Finally, as we have seen, the poet's interest in the Pythagorean vision of death and rebirth suggests a vision of historical time that only further highlights the intimate link between past and present. Rather than using the descent as a motif for dramatizing moral repentance, he uses it to represent a direct metaphysical connection between the Trojan Eneas and his descendants and ascendants, between the Eneas anchored in the past and his predetermined future destiny.

Yet some decisive shift must be imagined in order to account for the legitimate courtship of Eneas and Lavine and to set this episode and its conjugal success apart from the tragic events surrounding Dido. To account for this apparent polarity, Poirion makes a singularly unconvincing assertion that leads back to the moralistic viewpoint he, largely alone among critics, tries to refute. He argues that Dido's love is presided over by Venus, while love in the closing episode represents a different agency: *Cupidon-Amor*. In the end he finds no avenue for resolving the idealization of Eneas as a founding ancestor on the one hand and Venus's patronage on the other.[52]

Only most recently has Christopher Baswell explicitly argued in favor of a new direction which de-emphasizes the lyrical motifs in order to highlight the preoccupation with patriarchal, martial, and imperial themes in the

Enéas. At the same time, however, Baswell is forced to distance himself from Patterson's view that the poem largely flattens Virgil's darker ironies in order to create a streamlined and transparent narrative that unequivocally promotes a doctrine of Norman manifest destiny. In order to make his point, Baswell alludes to an example from a lesser-known corpus of poetry on the Trojan story. In this example, the Dido scene has simply been omitted altogether, so that the "counter-tradition" that haunts Æneas's fame throughout the Middle Ages is very purposefully erased. If such radical options were available, then one must try to account for the number of competing perspectives generated by the *Enéas* poet. In more simple terms, why would the poet who seeks to make the Virgilian epic into a work of Norman propaganda "complicate" its ancestral hero and the destiny of its ancestral race by dwelling on his amorous adventures? At this point, however, Baswell proves slightly evasive; he seeks an answer not in intellectual or doctrinal terms, but in terms of literary genre. He notes that critics have pointed to the importance of deferral and suspense in the plot structure of the medieval romance and argues that a form of acquisitive desire linked to the feminine and to feminine excess serves to retard the narrative progress and threatens historical progress, as if the poem's underlying moral posture were dictated by the requirements of the genre, almost against the poet's better judgment. In other words, by starting with the proposition that the *Enéas* poet is motivated by purely secular Norman interests, Baswell is forced to explain retrospectively everything that seems to cast doubt on the truly *manifest* nature of the Norman triumph.[53] Baswell also never addresses the work's mythological features, Venus in particular. Yet, it is hard to disassociate her from Eneas's extensive quest. One can only parcel the narrative by dividing love into good and bad forms, exemplified through sinful (unreformed) and virtuous (reformed) characters.

Venus's ubiquitous role is, however, the element which unifies the narrative. We have seen that Baswell touched upon this contradiction when observing that the same forces of amorous desire which threaten to defer or obstruct Eneas's destiny suddenly become, upon Lavine's appearance, a central feature of its accomplishment. At this point Baswell's analysis provides an unconscious glimpse at the narrative's overarching typological unity—a unity at once galvanized by the interventions of Venus and emblematized (typologically) by the goddess's inherent connection with the Trojan race—with Paris and Eneas as illustrious lovers. Indeed, we should reiterate that the *Enéas* poet seeks not to erase that medieval memory. For even if Dido has an integral and indispensable role in the narrative proper, there is certainly no compelling need to dwell on yet another level of even more problematic ascendancy going back to Paris and the abduction of Helen. Yet, the *Enéas* poet preserves Virgil's goddess and gives her a prominent role, just as he evokes Paris's name (near the beginning and end of Eneas' own itinerary) and begins the narrative with a reference to

Menelaus and Helen's abduction. Finally, the poet ascribes speeches to the queen that go a long way in recalling the broad outlines of the rich counter-tradition that taints Eneas's reputation.[54] Thus while Baswell's lucid and close reading of the work succeeds in identifying this tension and in placing the emphasis back on epic and historical themes broadly speaking, he falls back into a reductive notion of moral exegesis in order to *get through* the Dido episode.

Certainly one can look at Dido as an example of misfortune, but the closer one looks the harder it becomes to align the vocabulary of "excess" that surrounds her character with amorous sins. She loves (passively suffers love) no differently than do Eneas and Lavine. The presence and intervention of Venus throughout the episode places the depiction of love on a different plane.[55]

Thus it is difficult to accept Baswell's claim that the fundamental "tension between Eros and history that lies at the center of . . . the 'romance Aeneid,' particularly as it enters the vernacular literature of the high and later Middle Ages" (185) also lies at the center of the *Enéas* as he suggests. On the contrary, by incorporating a number of parallel mythologies, classical, Celtic, and Christian, the *Enéas* author attempts to recuperate the mythico-historical status of the Trojan-Norman race by integrating *eros* (under the aegis of Troy's divine patroness) to the designs of historical providence and by explicating its importance as an agency of divine order through heroic fictions of love and marriage dramatizing an extensive cultural *translatio*. Once classical history is divided along racial lines (Greek vs. Trojan) as part of the Norman myth,[56] the twelfth-century writer finds himself, along with his public, poised to *uncover* in the legendary transgressions of Paris and Eneas not a mark of moral inferiority but a mark of distinction. But how do the parallel pagan mythologies in conjunction with the tenets of contemporary marriage doctrine allow them to accomplish such an inversion of values?

The dualistic approach is esthetically, intuitively, and critically compelling, but it assumes that the Ovidian tropes are charged with a considerable potential for poetic innovation and nuance. For a noble person to exhibit courtly as opposed to ignominious *behavior* is one thing; to experience different *kinds* of venereal love or different depths of secular amorous consciousness requires a psychological and semantic subtlety virtually precluded by both the static nature and the underlying spirit of Ovidian motifs. This would require something more akin to the sophisticated speculation and imagery found in theological treatises on love which provide a space for describing spiritual interiority and transcendence.[57]

If we look at the work within the context of twelfth-century historicism proper, however, other typological perspectives make themselves available. In essence the break from the theologically oriented "universal history" gave rise to two prominent alternatives: a model based on lineage (anticipated in the genealogical structures of the *chansons de geste* and exploited as a model for the cultivation of royal and aristocratic "histories") and a model based on the revival of antique science that envisions history from the perspective of "fate," "fortune," and cyclical repetition over time.[58] Whether more or less serious or fantastical, such (pseudo-)histories stressed *continuity* within change: the continuity of lineage and descent underlying chronological "progress"; the continuity of cyclical repetition underlying the transitory whims of Fortune; the hidden designs of Fate or Providence—an impersonal and supra-temporal order—underlying potentially irrational or inscrutable historical events and the rise and fall of powerful men and nations. How then can we logically, within the larger sphere of historical and human genealogy linking Troy with Latium, Paris with Eneas, divide characters from themselves (i. e. the "Dido" Eneas from the "Lavine" Eneas) or from legendary forebears when lineage informs and sanctions the reading and legitimizing value of history for secular patrons? How can we do this without undoing the very mythological underpinning of the myth?[59]

Marriage fictions play a key role in these delicate and subtle negotiations and our true appreciation of the marriage theme itself is intimately tied to historical, amorous and mythical perspectives. The narrative marriage plots we have examined so far, for example, reveal the enduring legacy of pagan sovereignty and sovereignty bride motifs. This pagan motif is evidence of the profound cultural belief in the ritual and transcendent significance of aristocratic marriage practices and in the intimate relation between marriage and the transfer of political dominion. This mythic motif had additional resonance for medieval authors in light of contemporary (twelfth-century) historiography and theology. For as Curtius has pointed out, medieval historical thought found in the Bible a Judaic vision of *translatio imperii* which provided further "theological substantiation for the replacement of one empire by another" (Curtius 28).[60] In the twelfth century, this Old Testament motif was revived in a new form due to the prominent role in historical events which, as a legacy of Boethius, would be attributed to the Roman goddess *fortuna* who is identified as an important agency of fate and history (Patch 59–60; 113–4).[61] Moreover, Venus's own crucial role in governing Æneas's destiny would provide a mythical basis for rethinking the importance of love (over arms) as a guiding force in the transfer of dominion, and Virgilian and Ovidian influences probably played an important role in what Patch identifies as an enduring correspondence between the agencies of *Fortuna* and *Amor*.[62]

Of course, beneath the new historiographical and antique interests, there survived the overarching theological preoccupation. The historical survey of political dominion was motivated by the search for signs of a new divine covenant. At the center of this theological-historical nexus, therefore, we find a tension between orthodox and secular perspectives similar to the one we have been exploring in twelfth century narratives and in critical approaches to love and marriage motifs. Curtius illustrates this crucial debate with reference to the intellectual divide that separated Augustine's vision of Rome from that of Dante. For Augustine, the secular virtues of classical Roman culture are ideals tainted by sensual sin, and true believers should turn their attention away from Rome and toward the City of God. On the other hand, Dante's synthesis of historical and moral typology radically reoriented this view of Rome's relation to the central political and religious supremacy of papal Rome (Curtius 29–30). By focusing less on Augustinian dichotomies and emphasizing the significant role of sovereignty myths and marriage plots within the context of political history, we can begin to cast new critical light on these early and influential *romans* and more readily appreciate the links between theological and social perspectives in what are ostensibly works of secular interest.

It might still be objected that the process of "rationalization" which is typically seen to be a feature of the *romans d'antiquité* and the concomitant elimination of mythological elements would seem not only to conflict with our reading but to go hand in hand with twelfth-century interests in moral psychology.[63] Jodogne has argued, for example, that the interest in individual judgment is an innovative feature of the *romans d'antiquité* which influences the structure of narrative causation:

> On le sait, l'auteur de l'*Enéas*, comme l'écrit fort bien Pauphilet a supprimé ou fortement abrégé la mythologie, la religion et l'histoire, c'est-à-dire tout ce qui fait de l'*Enéide* un poème national. Et c'est ainsi que ce n'est pas Junon qui intervient auprès d'Amata pour qu'elle s'oppose au mariage envisagé par Latinus entre Lavine et Enée (VII. 286–358). C'est dans L'*Enéas*, *motu proprio*, la reine qui a "dessentu" son mari (3281-4). Selon Virgile, c'est Allecto la *luctifica*, la "faiseuse de deuils", qui avertit elle-même Turnus du risque qu'il court de se voir évincé dans son projet de mariage (VII. 406–34). Dans *Enéas*, c'est la reine qui prévient le Rutule (3385–98). Et la reine intervient ailleurs avec plus d'insistance. Dans ses reproches à Latinus, elle lui rappelle qu'Enée a fait mourir Didon de désespoir (3309–13). . . . Le rôle des hommes est plus en valeur, du fait que tant de déesses vengeresses sont éloignées. . . . les hommes décident davantage de leur destin . . . (79)

> [*We know (as Pauphilet has clearly demonstrated) that the author of the* Eneas *suppressed or considerably abridged elements relating to mythology, religion, and history—in other words, everything that made the* Aeneid *a national epic. Thus, it is not Juno who intervenes so that Amata resists the marriage between Lavinia and* Eneas *that is envisioned by Latinus (VII. 286-358). In the* Eneas, *it is the queen who,* motu propio, *disapproves (3281-4). According to Virgil, it is Allecto the* luctifica, *the maker of mourning, who herself warns Turnus of the risks involved should he be denied his marital ambitions (VII. 406-34). In the* Eneas, *it is the queen who warns him (3385-98). Also, the queen intervenes elsewhere more*]

forcefully. When she reproaches Latinus, she reminds him that Eneas had led Dido to die from despair (3309-13) . . . . The role of the human agents has more significance, since numerous vengeful goddesses are distanced from the action . . . . Human agents play a greater role in shaping their own destiny.]

We have begun to observe, however, that many of the actors in the *Enéas* do not necessarily "décident . . . de leur destin" as Jodogne contends. Throughout the narrative, Eneas's course is sanctioned by divine will and oracular prophecy: in the Dido episode, in the account of the Judgment of Paris, in the case of the final marital conflict, in all of these, the author of the *Enéas* goes out of his way to diminish the role of individual motivation and to give precedence to the "will of the gods." Eneas learns that he must simply leave Carthage, and the poet *repeatedly* invokes the will and "providence of the gods" (vv. 1617; 1623–24; 1634; 1639; 1657). When explaining to Dido the reason for his departure, Eneas insists upon the duties imposed by the gods:

> — Sire, por coi me fuiez donc?
> — Ce n'est par moi. — Et par cui donc?
> — C'est par les deus, qu'il m'ont mandé,
> qui ont sorti et destiné,
> an Lonbardie an doi aler
> iluec doi Troie restorer.
> Ensi l'ont dit et destiné;
> car, se ce fust ma volenté . . .
> n'alasse oan de cest païs
> Ne fust la volanté as deus . . .
> A enviz faiz la departie,
> ne est par moi, nel quidiez mie . . . (vv. 1757–64; 1768–9; 1775–6)

> [*"Lord, why then do you leave me?"* —*"It is not on my own account."* —
> *"Because of whom then?"* —*"It is because of the gods, for they have ordered me, having decreed and predestined that I must go to Lombardy where I am to restore Troy. This they have ordered. For, if I had my way . . . . I would not presently be leaving this country, if it were not for the will of the gods . . . . It is with regret that I depart. It is not because I want to; never think it so . . . . "*]

In the same speech, Eneas also implies that his own destiny is part of a larger historical order determining past events as well (vv. 1769–73). Eneas's words are corroborated by the poet's own allusions to the preordained operations of destiny (vv. 525–7; 574–80).[64] Indeed, the indulgent sentiments attached to Dido are not merely reflective of a Christian stroke of the brush but are part of a tragic sentiment perfectly consistent with the poet's dominant perspective. The author of the *Enéas* calls attention to the passive aspect of his heroes' amorous fate by electing to maintain the device of Venus's love spell (vv. 764 ff.). This device, like the love potion described in Béroul's *Tristan*, complicates and hinders any rush to moral judgment.

The poet figuratively but effectively epitomizes the real cause for Dido's downfall when he observes that nothing could threaten her amazing fortress (497–504)—"ne engin ne li forfeïst,/ *se devers lo ciel ne venist.*" As Poirion (1976) points out, the *Enéas* poet invites us to see Dido's downfall in terms not (only) of her private conduct but rather in relation to a struggle between larger forces vaguely adumbrated by the figures of Juno and Venus (215).

In the realms of providence and *eros* therefore, the poet consistently allows himself recourse to the incorporation of mythological divinities and agencies despite his tendency to pass over many of Virgil's mythical, religious and grotesque allusions. Although more limited in its scope, mythology clearly remains an active feature of the *Enéas*, and the poet's handling of the Judgment of Paris (vv. 93–182) is revealing in this regard.

Ostensibly, the Judgment of Paris provides a key link in the chain of illicit acts of love leading up to Eneas's own star-crossed encounter with Dido. It is interesting to note, therefore, that the poet does little to "rationalize" the fabulous features of this mythological vignette as he likely found them in his sources.[65] Unlike Benoit who will develop the character of the two famous lovers, the *Enéas*-poet leaves them one-dimensional. Neither Helen nor Paris is given a significant "role". It is the goddesses who provoke Paris, who define his choices and who seek revenge afterwards. The whole episode serves on an immediate level as an historical allusion not a moral analogy: the story is used to explain how Eneas's itinerary is determined by Juno's lingering anger at the Trojan prince Paris.[66] In this way it foreshadows the war between Paris and Turnus that will erupt over claims to Lavine and underscores the "Trojan" signature attached to Eneas's own destiny. The interest in the moral motivations behind the marital conflict is displaced by the mythical agency of autonomous pagan deities (and abstract personification in the case of *Discord*).

At the same time, the reference to the past event serves to highlight the ways in which the figure of Paris *prefigures* that of Eneas. Both Paris and Eneas will carry out through their actions an ostensibly *illegitimate* usurpation of dominion: Paris *vis-à-vis* the Greeks, Eneas *vis-à-vis* the Latins. Both will do so under the aegis of Venus and by means of a marital usurpation.[67]

We now see that the Eneas who marries Lavine does not necessarily represent a radical departure either from an established pattern of historical repetition or from a prior personal history. The alliance of love and dominion, the close connection between marital and imperial rivalry provide constant features of the poem's vision of manifest destiny. Nowhere, perhaps, is this underlying mythological design more evident than in the dramatization of the "bride quest" leading to Eneas's marriage with Lavine. In this (the second) part of the narrative, the queen plays a significant role,[68] and it is the queen who, in Jodogne's assessment, provided a particularly salient example of the "rational" perspective operating in the *Enéas* (cf. Poirion, 1976 216). Yet, while it is indeed true that the *Enéas* poet "rationalizes" the queen's character (Virgil depicts Amata as a raving bacchic fury goaded by a purely irrational passion) her relatively cogent arguments are in fact designed to underscore by contrast the insuperable dictates of an *absolutist* and *arbitrary* Providence.

The medieval poet's intentions in this regard are confirmed by his handling of Virgil's Latinus. In the classical model Latinus's unconditional support of the foreign suitor is the result of a supernatural and unmistakable augury (VII. 55ff.), and the ritualistic primitivism of his superstitious *prise de conscience* is attenuated only by the equally supernatural yet more destructive inspirations that motivate Amata's rejection of Æneas. Even in the purportedly more rational universe of the *Enéas*, however, Latinus is given no less obscure a footing: he only defends his position by a vague allusion to "the will of the gods." Thus, the extensive "rationalization" of the queen's arguments against the Trojan marriage alliance only *adds* to the apparently arbitrary nature of the king's divinely inspired defense of the marriage.

The importance given to divine intervention raises fundamental questions, however, about the role of love in relation to the marriage theme. If divine intervention is arbitrary and unprincipled, how are we to understand the metaphorical or symbolic meaning of the love and marriage uniting Eneas and Lavine in the context of a mythical fable of imperial manifest destiny? It would seem that Æneas, upon parting company with Dido, might have said not only "Italiam non sponte sequor" (IV. 361), but just as aptly "Italiam, non *sponsam*, sequor." With what end then, did the medieval author integrate the Ovidian stock to a (hi)story of foundation? How does he imagine the Trojan-Norman continuum, both in terms of racial unity and in terms of cultural inheritance and distance? What role do love and marriage play in this larger vision?

It is worth noting that Eneas's courtship ostensibly adds little to the course of the poem's action. Eneas is already pursuing the dictates of fate long *before* experiencing *amor*, and Yunck has observed (32–3, 37–8, 259–60) that the amorous interlude between Eneas and Lavine comes remarkably late in the narrative. Eneas's conquest in Latium and his claims to Lavine herself are ratified in prophecy long before any episode of amorous seduction ever takes place.[69] When Eneas initially learns of his important destiny from Anchises, love itself is insignificant. Eneas is a dynastic hero, not an amorous one. The emphasis Anchises places on geographical dominion and genealogical descent demonstrates the predominantly historical and imperial significance of Eneas's quest (vv. 2923–97).[70]

Finally, then, how does Ovid find his place in this absolutist myth of manifest destiny? To answer this, we need first to imagine how the medieval poet might have envisioned and conceptualized Venus's central typological features in relation to the Trojan race and the Norman myth of Trojan ascendancy. From this perspective, the pagan goddess appears not as the impetus for a cautionary moral *exemplum* but as a fitting patroness for an urbane, twelfth-century court aristocracy. If the medieval poet sought retrospectively to give a certain Norman patina to the Trojan saga, who if not Venus (the goddess of love) would provide a perfectly expedient device for doing so—a device that readily allows the poet to mirror an elitist ethic whose innovative characteristic was its cultivation not of war but of civility, manners, and *mesura* in accord with the dictates of *fin'amor*?

We know, and Baswell's erudite study of Virgilian traditions vastly adds to our knowledge in this domain, 1) that twelfth-century thinkers were perfectly capable of seamlessly syncretizing pagan allegory with Christian revelation, 2) that Virgil's cosmos was readily and frequently assimilated to that of Boethius[71] and 3) that Boethius's neo-platonic vision stressed the primacy of a transcendent providence over the localized agencies of moral volition in determining the fate of individuals throughout time . We know that Boethius' cosmos retained vestiges of Greek doctrines on creation and that the *Consolatio* perpetuated the concept of *eros* as an agency of order and progress at the micro- and macro-cosmic levels (*Consolatio* II. metr. viii. vv. 13–15; 28–30: "Hanc rerum seriem ligat / Terras ac pelagus regens / Et caelo imperitans amor . . . O felix hominum genus, / Si vestros animos amor / Quo caelum regitur regat"). Finally, we know that Boethian ideas were widely integrated with Christian thought as early as the tenth century (through the seminal commentary of Remigius of Auxerre; Baswell 126–7) and that this Christianizing, when it comes to prevail, does not inhibit commentators from continuing to expand upon thematic parallels between Boethius and the *Aeneid* (Baswell 126–7).

What is Ovid's place then? Once *eros* finds its proper place within this larger cosmological schema, Ovid would serve as the master who casts new light upon the nature of *eros* in its manifestation at the level of human agency and experience, especially with regard to heroic figures who have set in motion important and memorialized historical events. Although Ovidian tropes could be used to other ends, ironic, satirical and moral as well, his place within the present configuration can readily be defined with reference to another popular twelfth-century model, that of the macro- and microcosm which served to explain the correspondences and linkages between two orders, the providential and the individual (Chenu). Although secular in tone and vision and not cosmological, the *Enéas* poet would easily find ready concepts for envisioning Venus's central role in a myth of racial and imperial dominion in a way which gives Christian (neo-platonic) meaning to the lyrical tropes of Ovid, who was likewise considered by medieval thinkers as a *magister* in his own right. This convergence of pagan mythology and affective physiognomy is very neatly adumbrated in the conceptual approach to glossing the *Aeneid* defined by Bernardus Silvestris.

Although Bernardus will develop a different focus (this must be stressed), his concepts, drawing on the dynamic view of man as microcosm, provide an immediate locus for understanding amorous suffering within the context of private fortune and universal providence:

> Nunc vero hec eadem circa philosophicam veritatem videamus. Scribit [Virgil] ergo in quantum est philosophus humane vite naturam. Modus agendi talis est: in integumento describit quid agat vel quid paciatur humanus spiritus in humano corpore temporaliter positus. . . . (3 ll. 8–11)
>
> [*Now in truth we see this very thing as regards philosophical truth. In his role as philosopher, [Virgil] describes the nature of human existence. His method is as follows: allegorically he depicts what the human spirit does and suffers while temporarily trapped inside the physical body. . . .*]

At the macrocosmic level, twelfth-century writers might readily understand Venus not in association with *cupiditas* or vices *per se*, but in terms of the Holy Spirit. By the twelfth century, a Christian conception of *eros* was fused onto Boethian thought alongside the explicit notion of cosmic love alluded to by Boethius himself. For, with the commentaries of William of Conches, the World Soul adumbrated in the most famous passage of the *Consolatio* (III. metr. 9) is glossed as divine love:

> L'Ame du monde est pour lui [Guillaume de Conches] l'Esprit-Saint, c'est-à-dire l'amour divin qui produit la vie et comble de dons nos âmes individuelles; il repousse avec horreur l'opinion des vieux commentaires qui assimilaient cette Ame du monde au soleil vivifiant. (Courcelle 310).[72]
>
> [*For him [William of Conches] the World Soul is the Holy Spirit, i.e., divine love, from which all life derives and which fills our individual souls with a profusion of gifts. He is horrified by the old commentaries that saw the World Soul as being the vivifying sun.*]

Finally, William like Bernardus, conceptualizes man's place in the universe by reference to the macro- and the microcosm: the World Soul stands in immediate relation to the individual soul. Pierre Duhem points out that William largely aligns himself with "la doctrine monopsychiste" in asserting that "en l'homme donc, il y a une âme propre et l'Ame du monde" and in defining the World Soul as "une substance incorporelle qui est tout entière en chacun des corps" (96).[73]

At the same time, the poet would find that the Ovidian tropes, which depicted love in terms of passion—as passive suffering (*passio*)—would conveniently echo the features of stoicism central to the *Aeneid* and perpetuated in Boethian thought.[74] Accordingly, the lyric sequences that culminate in the marriage between Lavine and Eneas do indeed represent little more than an amplification, at the microcosmic level, of the poem's more sweeping and predetermined historical drama. In terms of form too, the mixing of lyric and martial motifs may seem artificial. On the other hand, philosophically speaking, it could be said that the poet rather creatively and innovatively fuses Ovidian lyricism and sentimentality with Virgil's stoic attitudes and themes (fatalism, alienation, and pious suffering and resignation). At certain junctures, Lavine's amorous suffering is cast in strikingly stoic terms and evokes the future hardships anticipated by Eneas from his perspective in Hades:

> se il i a un pou de mal [in love]
> li biens s'en suist tot par igal.
> Ris et joie vient de plorer,
> grant deport vienent de pasmer,
> baisier vienent de baaillier,
> anbracemenz vient de veillier,
> grant leece vient de sospir,
> fresche color vient de palir. . . .
> se il te velt un po navrer,
> bien te savra anprés saner. . . .
> [Lavine:] —Molt est ainçois chier comparé.
> — De quel chose? — De mal sofrir.
> — Molt estuet chier espanoïr
> lo bien, ainçois que l'an an ait. (vv. 7959–8011)

> [*If there is something of suffering [in love], happiness will surely follow. Tears will turn into laughing and joy; swooning is followed by sport and pleasure. After yawning come kisses, while embraces follow upon sleepless nights. In the wake of many sighs, comes much gaiety. Pallor is followed by renewed color . . . . If love wants to inflict harms, soon enough it will know how to heal . . . . [Lavine:] "One pays dearly beforehand."* —"In what way?" —"Suffering hardships." —"How costly to pay for a good before one has received it."*]

The *Enéas* adheres to the ethos of "piety" elaborated by Virgil: just as "Æneas submits, through love, and deliberately comes to serve fate,"[75] the medieval hero and heroine are enlisted in the service of Fate through the device of an allegorized *Amor* ("Bien doiz ester de as mesniee" 7993). Lavine, despite her good fortune, is no less a servant of Fate than Dido was before her. Her amorous passion provokes the same alienation: "voille ou non, amer l'estuet" (8061). Like the burden ("laborem") that Fate places on Æneas's shoulders (IV. 233), *Amor* enlists Lavine into the service of Fate against her will:

> Amor de menace n'a cure ,
> trop me moine grant aleüre;
> a ce que m'a chargié grant fais,
> lasser m'estuet, se ge nel lais;
> le sostenir et l'enchaucier,
> m'esmaie molt l'aprochier. (vv. 8439–44)

> [*Love heeds no dangers. Galloping forward at full speed, having inflicted so many burdens upon me, it must no doubt exhaust me if I do not let it go. Love's approach frightens me, so too the endurance and the persistence required.*]

It is in vain that the queen reminds Lavine that Turnus is devoted to her and that their courtship has endured seven years (8612–16). Lavine surrenders to *Amor* and *Fatum*:

> Quel deffanse ai ancontre amor?
> N'i valt noiant chastel ne tor,
> ne halt paliz ne grant fossé;
> soz ciel n'a cele fermeté
> qui se puisse vers lui tenir,
> ne son asalt gramment sofrir . . .
> Lo Troïen me fet amer . . .
> Quidez vos donc que bel me soit
> et que gel face de mon gré?
> C'est ancontre ma volanté. (vv. 8633–46)

> [*What defense have I against love? Castles and towers are powerless against it, no less any high palisade or broad moat. There is nothing under heaven firm enough to be able to withstand or endure love's assaults. . . . The Trojan man makes me feel love . . . .* Do you think this is agreeable to me, and that this is so by my own choice? On the contrary, it is against my will.]

Here too, as in the Dido episode, the author alludes to the supernatural agency of *Amor*. Lavine queries, "n'est Cupido frere Eneas, / li deus d'amor qui m'a conquise? / Vers son frere m'a molt esprise" (vv. 8630–32).

At that point where Ovidian convention dovetails with Virgilian mythology, therefore, the medieval author can be seen consciously to preserve an effective, transcendent, and anthropomorphic deity. This fictional device, however, allows for a unique blending of neo-platonic naturalism and Ovidian lyricism.[76] Specifically, the author adumbrates a Boethian notion of *eros* that allows him to synthesize providential and dynastic themes with the tenets of cultivated secular love. The operations of Providence and historical destiny are dramatized on the macrocosmic level in the figure of Venus, while Ovidian rhetoric provides a language for the exploration of *Amor* that integrates the individual experience with universal ideals.

Therefore, the significance of Eneas's final marriage must be understood in relation to the mythological determinations that frame the Latin conflict. The philosophical and mythical perspectives allow us to obviate the problematic "moral" aspects of Venus's associations with Eneas (while still affording room for subversive ironies). By virtue of this pagan device, the author of the *Enéas* is able to integrate his own theological and metaphysical speculations along with the nascent appreciation of Ovid into the martial and imperial themes of epic legend.[77]

In the second part of the narrative, the poet more sharply delineates the competing registers of logic governing Eneas's final conquest. As the marital struggle unfolds, the poet shapes the conflict between the rights of Turnus and those of Eneas according to the competing notions of *lex* and *fas*. To understand the broader significance of Eneas's bride quest and conquest, it is imperative to appreciate the poet's conscious attempts to underscore the tension between these two competing registers of perception and judgment—the rational and the providential, the contingent and the divine. Through the queen's warnings, the author of the *Enéas* purposely sets in relief the parallels between Dido and Lavine and provides a number of arguments that, within the immediate horizon of experience, would certainly seem to justify the expulsion of the foreign suitor.

When we are reminded by Turnus of his rights and of the honors the king owes him, we are forced to acknowledge the glaring illegitimacy of Eneas's claims. We know from the queen that Turnus has been courting Lavine for seven years already (8612–16), but just as love alienates Lavine from both her mother and her suitor, so too does Venus's intervention despotically and abruptly alienate Turnus's legitimate claims. Turnus reminds Latinus that his inheritance has already been granted and the marriage all but formally consummated:

> de ta terre m'as erité,
> o ta fille m'as tot doné;
> ge l'ai ancor pas esposee,
> et ne geümes an un lit,
> encor remaint par ton respit;
> mais de la terre sui saisiz,
> et les chastiaus ai recoilliz,
> g'en ai les tors et les donjons
> et les homages des barons;
> se tu me vels de droit guenchir,
> cil ne me porront pas faillir,
> tuit se tandront o moi de bot. (vv. 3849–61)

> [*You bequeathed me your land and granted me all, along with your daughter. I have not yet married her, and we have not shared the wedding bed. I desist, respecting your delay. But I have seized the land and taken possession of the castles as well as the fortifications and towers; furthermore, your barons have paid me their homage. If you think to deny me my due, they will surely not fail me, and they will side with me.*]

Here, for a moment, the marriage conflict takes a somewhat "realistic" turn. On the one hand, Turnus's speech makes allusion to twelfth-century preoccupation with the formal requirements of marital consummation. It suffices here to point out that Turnus's claim to Lavine has, legally speaking, not yet been formally actualized since there has been no formal exchange of vows and, as he says, no act of physical consummation. At the same time, we see the poet describing what Duby would call a marriage based on the lay model of prerogatives. In other words, Turnus's claim is based on his feudal relation to Latinus and his rank or *honor* defined in terms of his relative land holdings.

Moreover, the urgency of the claim stems from the transfer of land and wealth it involves. If Turnus and the queen share similar arguments and rationales for contesting the foreign intrusion, they implicitly share similar economic interests. Just as Turnus stands to lose his inheritance and power of dominion, the queen is no doubt conscious of the elderly king's impending death and of her own rights to a widow's inheritance. Should those lands be usurped by foreign intruders, her own future will be at risk. The opposition between Turnus's customary rights and Eneas's triumph seems intended to evoke the tension between a local feudal baronage and an imperial aggressor and may be intended as a reflection upon tension arising from encroachments on land and rights made by powerful Norman-Angevin rulers.[78] This contemporary dynamic is glimpsed in Turnus's unconventional (and anachronistic) description of the Trojan threat: the Trojans—traditionally depicted as the victims of defeat and subjugation—are now presented as being the usurpers of power much like the Anglo-Norman must have been perceived by many English barons (vv. 3823–42).[79]

The contest for Lavine's hand dramatizes the importance of marriage and of authority over marital grants in the struggle for dominion in medieval feudal society. Painter has pointed out, for instance, that in twelfth-century England the extension of feudal institutions gradually took legal power away from larger family units and placed it in the hands of lords and their immediate heirs (196). This image of shifting power seems to be reflected in the autonomy of Latinus's authority and in the standoff that takes place between Latinus and Turnus who is destined to become Latinus's immediate male heir through marriage to Lavine. It is perhaps to this realistic register that Poirion refers when he emphasizes, along with other critics, the medieval author's rationalizing tendencies: "L'économie est ainsi faite de tous les prodiges et de l'oracle qui empêchent, dans Virgile, le mariage de Turnus. La notion de destinée se prêtera mieux à une interprétation rationelle et moralisante de l'action du héros . . ." (216).

Yet this realism actually serves to delineate the poem's mythological subtext and to highlight the poet's commentary on the fallacy at the center of Norman efforts to legitimize their imperial claims. Latinus's late and arbitrary rejection of Turnus would clearly be an affront to customary prerogatives inherent in a well-developed feudal hierarchy of rights, and Baswell's contention that Turnus represents values "newly archaic" seems not fully to appreciate the extent to which the poet simultaneously deprives the emerging system of values of any identifiable and "realistic" rationale. Baswell says that Latinus and Eneas make claim to a new system of inheritance based on lineage and unilateral descent (209–8), but he fails to consider the significance of what he implies—namely that the text provides no *explicit* rationale for any such norm. There is no fictional *porte-parole* to counter Turnus's view. This system of "values" is dramatized more precisely as an *absolutist* determination, expressed prophetically and metaphysically. The absence of any normative rationale serves to highlight the purely *mythical* nature of the devices which ratify Venus's supremacy.[80]

Indeed, when looked at through the prism of social realism, love proves to operate at a level that transcends reason. In essence, the abrupt alienation of Turnus from his customary and normative right to make a claim on the wealthiest heiress in the realm is not only patently unjust, but flagrantly unrealistic. Speaking again with regard to evidence from twelfth-century English practices, Painter points out that "royal favor" was the one means for acquiring property whose potential surpassed that offered by marital strategies (200–203). And, in some cases, the king's authority to preside over marital grants was "absolute" (206). Yet, when Painter explains that marriage was, in any event, governed by economic interests, the patently fantastical nature of Eneas's happy destiny is set in relief (206: "Rarely was a man in such high favor that he could secure an heiress in the king's gift *without paying heavily* for her," my emphasis; 208: "If a younger son was to be married to a lady of high position, *the girl's father would insist that he have a fief of his own*," my emphasis). The queen derisively reminds Latinus that Eneas's claims to the king's daughter are not backed by any real wealth or "brideprice" (vv. 3299–3330; 3367–71). Her remarks not only reflect upon underlying economic tensions but highlight the purely mythical and providential aspects of Eneas's challenge.

Thus the glimpses of contemporary political conflict revealed in the marriage rivalry are punctuated by a renewed reference to the purely mythological "logic" that subtends imperial authority. For while it may be true that the twelfth century saw the "rise of patrilinear inheritance" (Baswell 208), it also remains true that from Turnus's and the queen's immediate "horizon of perception" (I am evoking a Boethian schema), the marriage of Eneas to Lavine could only appear as a total *alienation* of land and power running counter to *any* perceivable logic and precedent. This, again, is why the queen is fashioned as being the most "rational"—in defending her interests she serves in the larger design of the work to expose the illegitimacy underlying imperial impositions, to expose the importance of the surviving mythological residue that continues to operate at the very center of the work, at the center of Norman historiography.

Whatever the political tensions adumbrated in the Turnus-Æneas conflict, the triumphant conquest of the Trojan prince conveys in the end a striking disregard for mimetic cogency. In underscoring the providential nature of Eneas's victories, the poet seems bent on reminding us of the very "mythology" underlying imperial claims to dominion and, by extension, imperial ideology and propaganda. It is not so much reasoned principles but rather the evident futility of Turnus's resistance that leads Drances and other barons to abandon his cause. But the fortunes of Eneas, like the misfortunes of Turnus, are emphatically put forth as being, from the viewpoint of reason and moral providence, as arbitrary as they are ineluctable. Ironically, therefore, history would indeed seem to be repeating itself.

Given Turnus's well- and long-established claim to Lavine, Eneas's intrusion and illicit act of abduction must needs be (from Turnus's perspective at least) reminiscent of Paris's own famous crime. This leads, of course, to yet another irony: it projects an unexpected re-valuation backward through time onto Paris's own actions. The Boethian vision consistently provides a model for these divergent perspectives (the limited, contingent and "moral" judgment vs. the inclusive and providential), and regardless of which perspective one adopts, it must be projected along an unbroken diachronic line: one cannot divide the narrative or the genealogical and historical continuum.[81]

In the end the text refers to the practice of unilateral descent only very obliquely, as a future reality and as an inherent aspect of the unbroken line of descent that is essential to the myth of the Trojan-Norman continuum. In no other way does this shift in secular values logically legitimize Eneas's intrusion and conquest. Instead, the poet turns our gaze toward Venus's agency. The fact that providence renders completely ineffectual the kind of secular norms invoked by Turnus suggests a turning away from established lay customs and values. But how does Venus legitimize this conquest? What alternative set of norms does the poet set into relief?

Of course, we have already begun to lay the foundation for answering these questions which themselves are an outcome of our examination of the *Enéas* and the *Roman de Troie* up to this point. It has been necessary to view these two *romans d'antiquité* in a broad and inclusive manner in order to isolate the most strikingly innovative elements in these "transitional" works—works which added important new dimensions to the political themes and literary conventions of secular epic by dramatizing the struggle for sovereignty not only on the plane of martial confrontation but in terms of rather explicit amorous conquests and personnified pagan dieties. Specifically, the *roman d'antiquité* offered not only a dramatic vehicle for the explication of diverse mythical legacies, but it established a working if nascent literary model for harmonizing in typological, rhetorical and artistic terms a rich confluence of pagan, neo-Platonic and (secular) lyric traditions. Not always esthetically satisfying, this syncretic enterprise finds in the figure *Amor* a common denominator that can readily provide an astonishing degree of thematic unity. The purely allegorical and rotely lyrical aspects of *Amor* gain new currency in turn as psychological (lyrical and experiential) and secular (political and ethical) visions are integrated with legendary events of epic moment and mythic resonance. We have also begun to see that the Troy legend itself (whose own peripities seemed to incarnate the relation between myth, history, love and war) emerged as a kind of archetypal referent for the evocation, dramatization and appropriation of this rich cultural *translatio*. At the center of this legend that accounts for a momentous *translatio imperii* is an act not of arms but of love and marriage—a defiant, even criminal act of spousal abduction.

Perhaps the emerging medieval fascination with Paris' crime and the fall of Troy is ultimately due to the compelling features of the archaic myth itself, or perhaps this monumental reception can only be explained by reference to the additional impetus of twelfth-century dynastic claims. Whatever the case, Paris' fortune was clearly distinct from that of Achilles and Roland, and in place of a martial hero an amorous criminal takes center stage. Accordingly, the events radiating from the hero's action and at the center of epic struggle find their meaning in relation to amorous and matrimonial reversals whose problematic moral aspects were clearly capable of soliciting inventive responses from both medieval authors and their public. Likewise, the Norman myth—by virtue of its dynastic claim and the typology which validates this claim—necessarily calls, at least initially, for an ostensible inversion of moral and martial values, for the rehabilitation of a dubious ancestry, and for a literary fabulation that consciously departs from, but ultimately recuperates, essential ideological features of feudal epic.

This reorientation of the values that ratify secular dominion is accomplished by linking it to marriage as a sacrament whose sanctity and divine dimensions are prefigured in pagan allegories. This linkage will of course readily allow for the integration of amorous themes. Initially, however, it converges with the Indo-European motif of sovereignty and the sovereignty bride. This mythical subtext is a prominent if only implicit feature of the rivalry over Lavine. It is also evoked more directly and unmistakably, however, with regard to the dramatization of the Trojan's future imperial destiny as well (as we shall see). Finally, it gains its full meaning within the narrative by convergence with another supernatural vision of marriage, that of the Christian view of marriage as a sacred institution whose sacramental efficacy is ratified through the inner operations of mutual affection and mutual consent between husband and wife (our focus in chapter 3).

In the twelfth century the prologue to the *Lais* of Marie de France provides direct testimony not only to the growing popularity of Celtic legends, but also to the readiness of continental writers to apply to them exegetical modes of thinking. Archaic Indo-European beliefs about territorial goddesses and the ritualized relations between sovereigns and their brides are virtually ubiquitous in surviving Celtic myths and sagas, and the same set of Indo-European traditions informs the larger significance of Paris's abduction of Helen and the magnitude of the Greek response. Despite the Trojan defeat, therefore, the abduction could be said (retrospectively) to prefigure the eventual transfer of dominion from Greece.[82] It is this pagan subtext that accounts for the exceptional urgency which is given to the marital rivalry in the *Enéas* and which implicitly defines the nature of the threat posed by Eneas. According to the norms of a feudal aristocracy, for example, Eneas's rag-tag band of disenfranchised outsiders would hardly pose a credible threat to Turnus. In the context of Celtic lore, however, the transfer of dominion can be and often is dramatized as the usurpation of a woman by a rival and often disenfranchised but divinely elected suitor. Furthermore, this usurpation is likely to be imagined not in military terms but in terms of amorous pursuit and seduction.

Thus, in spite of Turnus's status and real power within the circle of prominent barons, it is, almost magically, Lavine who becomes the sole criterium in the struggle for dominion as her very person grows indistinguishable from the territory under dispute:

> Li rois oï que Turnus dist,
> d'ire et de maltalant fremist . . . .
> D'entre ses homes s'est levez,
> an sa chanbre s'an est antrez,
> ne mes que sol itant lor dist:
> qui plus peüst et plus feïst,
> et tot li mist el covenir
> ou del laissier o del tenir;
> *la feme eüst o tot la terre*
> *cil d'aus qui la porroit conquerre.* (vv. 3869–80)

[*The king listened to Turnus's words and trembled with anger and resentment . . . . He rises from amidst his vassals and enters his chamber. He says only this: to each his own and may the most able prevail. He leaves them free to either surrender the land or to defend it*—the one among them who could conquer her / it, would have the woman along with all the land.]

When the queen's messenger alerts Turnus to the fact that the king is already making promises to the Trojan intruder, this conflation of bride and land becomes increasingly evident:

> il li abandona sa terre,
> anvoia li chevaus de pris
> toz anselez trois cenz et dis;
> *ansorquetot sa fille a fenne*
> *li otroia o tot son regne.*
> Ce saches tu bien sanz dotance,
> ne tient pas ta covenance,
> de sa fille n'avras tu mie,
> se tu nen as molt bone aïe;
> mes ce te mande la raïne,
> *qui velt que aies la meschine,*
> *tote la terre et lo païs*
> que te porchaz par tes amis.
> Prent soldoiers, asanble gent,
> ne te tarder mes de noiant,
> lo Troïen coite de guerre,
> tant qu'il te guerpisse la terre, . . . (vv. 3418–34)

[*He handed over his land to him and sent to him all outfitted his more than three hundred finest horses. On top of all this, he granted him his daughter for a wife along with all of his domains. It must be obvious to*

*you by now that he is not keeping his pledge to you. You will never have
his daughter unless you can muster powerful supporters. However, the
queen, who wants you to be the one to get the girl, all the land, and the
country, urges you to draw upon the support of your friends to acquire
them. Gather together your men and hire soldiers; make no delay. Make
war on the Trojan until he surrenders to you the land . . . .]*

It is impossible to distinguish between bride (*meschine*), land (*tote la terre*)
and political dominion over the land (*lo païs*).[83]It is of course conceivable
that all three go hand in hand in the case of a royal marriage. In the *Enéas*,
however, Lavine's marriage represents more than a *transfer* of wealth and
power, it represents a most unlikely *conquest* to be carried out by a band
of vanquished and exiled foreigners as I have stated ("est novelment uns
hom venuz/ de çaus de Troie, des veincuz" vv. 3413–14; see also Turnus's
own expression of incredulity with regard to the threat: vv. 3468–72). The
king's wishes may be as peremptory as the outcome ordained by prophecy,
but when the queen exhorts Turnus to muster troops (see vv. 3429–34
quoted above), we are again reminded that the poet makes little effort to
demonstrate what support the king relies on to enforce his authority. On
the contrary, Latinus seems isolated from the queen and his highest barons
by virtue of his antagonistic and indefensible position. At one point, as
Turnus reflects on the threat posed by Eneas' intrusion, marriage and
dominion are bound almost tautologically, and struggle is reduced, by
metonymy, to an all-or-nothing fiction:

> Bien l'oï dire des l'autr'ier
> que Eneas ert arivez;
> mais icist plaiz m'estoit celez,
> *que Latinus li don't son regne*
> *ne Lavine sa fille a fenne.*
> Il m'en fait pieça le don,
> ne la perdrai mes sanz raison;
> *si cil la velt vers moi conquerre,*
> *ja jor n'aie feme ne terre,*
> *se ge vers lui ne la deffent. . . .*
> Li rois es vialz et toz defraiz . . . .
> Ne me deüst de droit guenchir . . . .
> Droit li estuet que il me face:
> *il m'en a pieça fait saisir,*
> *ne m'en puet mie retollir*
> *ne la terre ne la meschine. . . .*
> Se bataille velt Eneas,
> en moi ne remendra il pas;
> *se par lui per plain pié de terre,*
> *mal porrai donc autre conquerre.* (vv. 3458–67; 3473; 3476; 3480–3;
>   3487–90)[84]

[*Of course I am aware of Eneas's arrival, but this agreement was not
revealed to me, that Latinus promises him his kingdom as well as Lavine*

for his wife. *Only recently he had granted these to me. I will not lose them without just cause.* Though this intruder may wish to steal her from me, never will he get either the maiden or the land without my putting up a fight . . . . *The king is old and decrepit . . . . He has no right to break his pledge to me . . . . He owes me justice.* Only recently he gave me possession of the woman and the land; he has no right to take either away . . . . *If it is a battle Eneas is after, he will not have to wait on my account.* If I were to lose even a foot of land to the likes of him, it would be hard to imagine my conquering anyone.]

Rhetorically, the mimetic contingencies of Turnus's struggle are subordinated to the metaphorical relation between bride and dominion: the parallel aspects of Eneas's tripartite destiny—as warrior, dynastic ruler and lover-husband—are clearly meant to be seen in terms of their mythical conjunction and not according to secular (judicial) norms or moral legitimacy. It seems that myth and metaphor are not placed in static or critical (i. e., "exemplary") opposition to the narrative action (Patterson 171–2); rather the action itself is premised upon a dynamic mythological design which gives new meaning to Venus's patronage: if the quest for dominion parallels a quintessential Celtic bride quest, then Eneas's mythic ancestry makes him a supreme contender.

The poet's deliberate integration of the Celtic sovereignty motif becomes more evident in his imaginative elaboration of the stag hunt carried out by Eneas's son Ascanius. The poet's handling of the episode reveals the extent of the mythic subtext governing Eneas's bride quest and marriage. The stag hunt is a prominent Celtic motif dramatizing the divine election of an emerging young hero and king figure. The currency of this Celtic motif in twelfth-century tradition is attested to by the tale of *Guigemar* in the *Lais* of Marie de France. Nearly a doublet of the sovereignty bride motif, it seems natural that this motif would be grafted on to its Virgilian parallel. This is readily accomplished through the addition of certain "marvelous" touches.

Since Virgil's text already contains the essential outlines of the stag hunt, it is difficult to judge what "Celtic" and "twelfth-century" aspects the poet might be exploiting (The passage begins line 3525 in the *Enéas;* in the *Aeneid,* see VII. 475ff.). However, at least one critic of the *Enéas* has commented on the artistic similarities between this episode and similar scenes from twelfth-century narratives such as *Guigemar* and *Erec et Enide.*[85] Moreover, one of the author's imaginary extrapolations comes in the the image of a marvelous candelabra (vv. 3556–64).While the poet seems to be influenced, at least in part, by his knowledge of the *Roman de Thèbes* (Yunck 126, note 72), it is interesting that luminous tapers are a feature of the Otherworld, as in the case of the Fisher King's palace for example, and that two fine gold candelabras also figure as the prominent feature on board the fairy vessel that transports Marie's *Guigemar* to the Otherworld isle immediately following the opening stag hunt described in that work (*Guigemar*, vv. 183–6). Rachel Bromwich has argued that the Celtic Stag Hunt represents one of the most prominent narrative motifs to have been passed on to Breton soil from earlier Irish and Welsh tradition as part of the Celtic sovereignty motif (Bromwich "Dynastic Themes" 460).[86]

Ascanius's request for permission to hunt the stag (3565 ff.) as well as the medieval author's decision to consummate the hunt reflect two apparently original additions to Virgil's text.[87] Both of these details find a common source in an ancient Irish sovereignty legend involving the hunting, slaying, and cooking of a marvelous golden fawn.[88] The hunt, significantly, serves to distinguish a future sovereign from among a number of equally young peers. Accordingly, in the *Enéas*, it is the young Ascanius as future sovereign who, after getting permission to proceed, successfully consummates the hunt. The episode also seems to parallel an analogous set of events in the *Roman de Troie*: (Benoit's) Paris asks Priam for permission to leave for Greece to abduct Helen. In both cases permission is granted to a royal son of Trojan origins and the subsequent act of hunting or abduction is linked to an impending struggle for and eventual appropriation of sovereign dominion.[89]

In her work *Insular Romance*, Susan Crane has likewise pointed out that the sovereignty bride motif remained a central fictional device in the late twelfth and thirteenth-century native, insular romance tradition. In harmony with our own assertions about the poet's elaboration of amorous and marital themes in the *Enéas*, Crane points out that although the love theme "draws value from contemporary courtly ideals of beauty and passion" its "vital function" is subordinated to "the hero's determination to reestablish his patrimony, with military action and courtship as a means to that end" (Crane, 35). The fictional and transcendent idealism of the marriage-sovereignty motif is further attested to by the fact that Crane's discussion bears upon texts which envision baronial struggles for repossession and the claims of disenfranchised barons *against* the encroachments of imperial law. We can conclude that the marriage motif in the *Enéas* conveys a similar political idealism although it is modulated to reflect on the fictionally elaborated claims of imperial sovereigns. In this regard the marriage theme would resonate deeply with contemporary audiences. The landmark marriage of Henry II and Eleanor, itself of immeasurable political and social consequence, was moreover a spectacular enactment of a potent Norman strategy. The dramatic consolidation of power accomplished in 1152 had been prepared by the earlier marriage struggle between Henry I and Louis VI to gain Angevin support. Henry I triumphed when he wed his daughter Matilda, the widow of emperor Henry V, to Count Geoffrey of Anjou. According to the prominent Norman historian D. C. Douglas, "the implications of the marriage were indeed to be far reaching, for it detached Anjou from the French king and at the same time made possible the future union of Normandy, Maine, Anjou and Touraine under a successor of the Norman Kings of England" (Douglas, 1976, 71; cf. 16). Furthermore, Douglas has very emphatically emphasized the role of marital strategy as a fundamental tool of Norman expansion and dominion. Hence, the amorous impress that marks the Trojan ascendancy finds a contemporary cultural echo in the marital-political savvy of the Norman rulers who exploit the myth of Trojan origins with unprecedented vigor.[90]

At the same time, however, the marriage fiction seems to alienate Latinus's marital strategy from other epic values and places it in a system of norms neither predominantly feudal nor military. Thus the secular practice is ostensibly given a supra-secular and supernatural set of sanctions and rationales, such that marital conquest is depicted as betokening a providential order of manifest destiny that transcends the authority conferred by brute military subjugation and economic superiority—which were precisely the cynical realities limiting Norman claims to sovereign rule in England and motivating the elaboration of a more compelling absolutist mythology. In the following chapter we will examine how the pagan and neo-platonic mythologies are integrated through a glossing of epic marriage fictions into a Christian order of values in accord with emerging doctrines on the marriage sacrament.

Again, however, the appeal to myth ironically forces us to remember the underlying ironies of the Norman imperial fallacy and the "challenge" posed by competing views of Eneas and the Trojan race in history. By resorting to supernatural agencies and calling attention to its own idealistic vision of history, the *Enéas* fosters and retains a provocative ambiguity. As Patterson points out, the Anglo-Norman peace established under Henry II was itself hemmed in by episodes (past and future) of flagrant civil violence (180). It is indeed this kind of dark fatalism that informs the tragic cycle of war, despair and betrayal in the tale of Troy and the Greek homecoming, and it is within the larger context of the "matter of Troy" that the ambivalence attached to Æneas's character most clearly emerges (Baswell 18).

This "counter-tradition" was based on the early medieval pseudo-histories of Dares and Dictys. "Throughout the Middle Ages, Dares and Dictys were generally thought to be more historically accurate than Virgil . . . ." (Baswell 18 and 19). It is the second *roman d'antiquité* in our corpus, *Le Roman de Troie*, that more directly confronts the problematic elements of the larger Trojan matter. How did Benoit de Sainte-Maure adapt this pessimistic vision to the political optimism of powerful patrons? Did he attempt to reconcile the infamous marital crime of Paris at the heart of his narrative to the mythical vision of a heroic and divinely fostered Trojan ascendancy? How does he reshape the crucial episodes of marital transgression and marital conflict so central to the Trojan saga and to the western myth of *translatio imperii*?

We will see that Benoit's treatment of this matter reveals the exceptional elasticity of twelfth-century exegesis. Benoit's text will help us to appreciate the Christian "myths" (if we may be permitted to say so) that provide a crucial subtext for our final appreciation of the marriage fictions in the *Enéas*, and which Benoit exploits to rehabilitate troubling features of the Trojan heritage. In order to appreciate the transcendent values underlying the treatment of marriage dramas in these early *romans*, we should digress briefly from our study of Old French texts to examine relevant aspects of emerging theological ideas about the religious nature and sacramental value of marriage.

NOTES

1. There also exists a fragment of a *Roman d'Alexandre* in Occitan by Albéric de Pisançon from 1138.

2. See the overviews and analyses in Yunck's introduction and in Faral (*Recherches* 169ff; "Compte rendu" 100).

3. Of course the sources referred to here are respectively Virgil's epic and the Latin chronicles of Dares and Dictys. I will discuss the *Roman de Troie* and its sources in more detail in chapter three. For the sake of consistency, I will use the Old French spelling when referring to the characters in the *Roman d'Enéas* (*i.e.,* Eneas, Didon, Lavine). Alternately, I will use the Latin spelling (Æneas, Dido, Lavinia) only when referring to Virgil's characters specifically or to their roles as defined by Virgil's epic.

4. For additional approaches to defining the esthetic innovations of the *romans d'antiquité*, see Frappier (1964) and Jodogne.

5. E. g. Adler, Lumiansky, Laurie, Jones, Cormier, Nolan, Patterson, Burgwinkle, Baswell.

6. On love and marriage themes in the *chanson de geste*: Jacques De Caluwé, "L'amour et le mariage, moteurs seconds, dans la littérature épique française et occitane du XIIe siècle." *Love and Marriage in the Twelfth Century*, ed. Willy Van Hoecke and Andries Welkenhuysen (Louvain, Belgium: Leuven U P, 1981) 171–82.

7. Especially true with regard to the *Enéas*, which has generated a greater mass of critical commentary (e.g. Adler, Laurie, Jones, Cormier, Nolan). The most notable exceptions are (to some extent) Poirion (1976) and quite recently Baswell (1995 chapt. 5). Baswell's overarching contention that "Dido and Lavine, because they provide much of the overtly original material in the poem, have so interested critics that their centrality has been exaggerated" and that "critics have tended to ignore the narrative's competing and finally greater attention to themes of patriarchy: militant aristocracy, the foundation of empire, and the establishment of royal power through battle, public argument, law, and lineage" (200), largely anticipates my own attempt to look at love and marriage tropes within the context of the work's larger narrative contingencies and thematic objectives. It is telling in this respect that Baswell's own analysis (i. e. chapt. 5) rarely relies in any considerable degree upon earlier commentaries. But what about the relative (in)compatibility of these critical perspectives? I believe the critical divide, never fully articulated or bridged in Baswell's analysis (more on this later) will help us to define the author's view of love and marriage in the *Enéas*. Finally, an even greater appreciation of the *Enéas* will depend upon an adequate reading of Benoit's retelling of the Troy story

within a socio-literary context contemporaneous with that of the *Enéas*.

8. Jean-Charles Huchet, *Le roman médiéval* (Paris, 1984), p. 19. Cited in Nolan (158–9).

9. Translation of H. F. Stewart (Loeb edition, 1918, rpt. 1962).("At vero hic etiam nostris malis cumulus accedit, quod existimatio plurimorum non rerum merita sed fortunae spectat euentum tantum iudicat esse provisa quae felicitas commendaverit. Quo fit ut existimatio bona prima omnium deserat infelices. Qui nunc populi rumores, quam dissonae multiplicesque sententiae, piget reminisci. Hoc tantum dixerim ultimam esse adversae fortunae sarcinam, quod dum miseris aliquod crimen affingitur, quae perferunt meruisse creduntur.")

10. "Aux yeux des Romains, le mariage légitime possédera toujours une valeur quasi sacramentelle, inhérente à la nature des choses: il s'inscrit dans l'ordre du monde, et contre ce fait, les volontés individuelles ne peuvent rien" (Grimal 38).

11. Cf. Patterson's argument, 180–1.

12. We will return to this larger question of the significance of medieval marriage fictions in the context of *translatio imperii* and "cultural progress" in chapters three and four.

13. The same perspective is emphasized in Eneas's reminiscence of the war (vv. 859–86). Nolan makes a parallel observation but with divergent conclusions in mind: she sees the reminiscence in terms of an exemplary strategy on the part of the author who seeks to contrast the past sins of the Trojans (orbiting around the figure of Paris) with the redemption figured in Eneas's culminating triumph (Nolan 160–1).

14. Adler, Laurie, Huchet, Cormier, and Nolan share this overall perspective.

15. Cormier asserts that the Dido episode is "evil" and that Dido is condemned beyond salvation (141), while at the same time he describes Dido herself as the victim of an "ineluctable and inalterable fate" who retains in the *Enéas* the original aspect of Virgil's "optima Dido" (Bk. IV.). Cormier goes so far as to note the contradiction, but provides only a weak explanation: "It is ironic, too, that the Old French text maintains her munificence: here, in being generous, she is no doubt sharing, as any human might, in God's goodness by imitating his ultimate act of generosity in creating the universe" (124).

16. It is, of course, widely accepted that the *Enéas* and the *Roman de Troie* elaborate on the Norman claims to Trojan ancestry (cf. Eley, Douglas, Patterson, Yunck's introduction). Within this context, critics have also speculated that the landmark marriage of Henry Plantagenet and Eleanor of Aquitaine (1152) provides a contemporary reference point for the *Enéas* and *Roman de Troie* (more on this below). On the rich and ineluctable ironies of the Troy topos, see Birns. Poirion (1976 216) points out that the poet also connects Eneas to Dardanus (vv. 4711–18) so as to reinforce the idea that the Trojan pedigree manifests itself through an extensive historical "lineage." It might be noted that Ovid exploits this motif to his own ironic ends in the *Heroides* where tragic repetitions over time are attributed to the influences of genetic descent and affiliation. Phyllis angrily accuses Demophoon of being his father's son (i. e. abandoning his wife as Theseus had abandoned Ariadne). A similar motif appears (with multiple ironies) in other letters (e. g. V; VIII). The same notion of genealogical continuity appears in commentary traditions as well: "Interest in lineage, and the particular attention to a theme of lineal return through Teucer and Dardanus, is very widespread indeed in the commentaries and recurs regularly in the *Enéas*. (Such lineal return of course corresponds nicely to the

Normans' original claims to English kingship)" (Baswell 177).

17. It should be noted that Patterson's initial assumptions about the *Aeneid* and its legacy explicitly rule out any such resolution (160): "Put simply, the tension that at once animates and inhibits the *Aeneid* is a struggle between, on the one hand, a linear purposiveness that sees the past as a moment of failure to be redeemed by a magnificent future and, on the other, a commemorative idealism that sees it as instead a heroic origin to be emulated, a period of gigantic achievement that a belated future can never hope to replicate. . . And so too the medieval rewriting of Virgil becomes itself . . . a site of both emulation and exorcism, of slavish imitation coupled with decisive rejection." (I will argue that the tension created by competing authorities and traditions motivates, involves and leads to a more complex synthesis of perspectives.)

18. Hermione's letter provides a model for the logical problem encountered by glossators: Hermione is not commended for her legitimate love; moreover, she apparently suffers in spite of it (cf. Hexter 186). Nor is there any reason to assume a sagacious cleric and writer would conform to the pedestrian logic that characterizes the Ovidian glosses to begin with. Speaking of the tripartite scheme which divided Ovid's heroines into examples of either legitimate, illicit or foolish love, Hexter himself immediately observes:

"One must grant the simplicity and clarity of this scheme. Of course, it rides roughshod over the subtle differences among the heroines and the situations in which each finds herself. . . . Desire to make commendation of legitimate love the overarching aim of the entire collection is an added reason why the commentators paid so much attention to *Epistulae heroidum 1*: Penelope is the canonical example of legitimate love, and her letter is first. There are others in this category . . . As the list illustrates, however, a legitimate marriage is no guarantee of an admirable heroine, and the headnotes to each epistle in the commentary show that reference to this supposed aim of Ovid was made only sporadically" (157). Hexter elsewhere points out that the introductory glosses name four additional typological categories for approaching the letters (162–3).

19. A thread of Ovidian discourse that does not seem to have been lost upon medieval readers who in commentaries on the *Ars amatoria* gloss Medea as being an exemplary figure for demonstrating "that men deceive more often than women" (Hexter 55–6). (Cf. our discussion below of Dido and Benoit's treatment of the Jason and Medea story.)

20. Hexter remarks on the tendency to ascribe reactionary, moralist attitudes to all medieval *auctores*: "It is clear, however, that in the monasteries and in the schools within their walls there was not the antipathy to Eros with which contemporary imagination credits them, nor even the squeamishness many nineteenth- and twentieth-century scholars have shown . . . Any latent aversion to the topic was obviously negated, as the texts themselves indicate, by the value-free function these texts had: instruction . . ." (25).

21. Yunck stresses the same conclusion and shows how the amorous motifs from Ovid— "largely concerned with feminine love psychology"— can lead to awkward results when they are suddenly appended to an older, war-torn hero like Eneas (36–37). But Eneas does suffer from love as well and this forces us to ask how the poet expects us to discriminate, in ethical terms, between Dido on the one hand, Eneas and Lavinia on the other.

22. Clearly the reader is distanced from the Queen by the fate that hangs over her;

however, her claims do not necessarily lack in reason and legitimacy (a point which the poet exploits by design, as we will see). The "counter-tradition" that plagues Eneas also adds weight to her words and no doubt the author would be conscious of such implications. Burgwinkle has shown how contemporary echoes also provide immediacy and additional resonance to the incriminations.

23. Later, Eneas' sincerity and respect are revealed in his esteem for Dido's ring and her robe as his most cherished gifts, vv. 3133–38; 6121–4. Again, for the feelings of anguish and dismay that Eneas experiences because of his mandatory departure: vv. 1625; 1629; 1631–2; 1635–6; 1640.

24. Rather than "progressing" toward "*mesure*" as the logic of refined love, mystical love or *fin'amors* would suppose (e. g. Laurie, Cormier, Poirion 1976, Baswell), Lavine's pronouncements grow increasingly severe: first she sees no recourse but death should she lose Eneas (vv. 8255–6; 8332–3); later she speaks of actively taking her own life vv. 8745–7); and finally, just before Eneas' combat with Turnus, she experiences her most unequivocal suicide threat: "angoisse est, grant peor a; / an son corage a esgardé/ et bien fermement proposé / si Eneas i est ocis / o par son enemi conquis, / qu'el se laira por soe amor / cheor aval jus de la tor: / ja anprés lui ne vivra ore." (vv. 9318–9325).

25. On various castings of the hunt motif, see Marcelle Thiébaux, *The Stag of Love: The Chase in Medieval Literature.* (Ithaca and London: Cornell U P, 1974). The *Enéas* poet will, however, modify and exploit in significant fashion a later hunting episode involving Ascanius's killing of the stag. More on this later.

26. But, as Jones points out, "in this case Dido's vow of faithfulness to Sychaeus takes precedence over such a consideration (39–40)." Why? An important point to which we will return later.

27. Threats identified by Baswell as 1) a rising mercantile class (exemplified by Dido and Carthage), 2) feminine sexuality (which converges metaphorically with mercantile greed and luxury) and 3) a feudal aristocracy determined to preserve its own powers, prerogatives and customs (Turnus and Drances).

28. For a critique of Robertson's rigid definition of the Augustinian dichotomy *caritas / cupiditas* and for a more fluid and dynamic conception, see Spence (97). On the role of desire in Christian rhetoric and in typological dynamics, see again Spence (chaps. 2 and 3).

29. A topos which reappears later as an important motif in *La mort le roi Artu*. In the *Enéas*, the influence may also be Ovidian. In the case of the *Heroides*, it is, however, typically the woman who imagines an epitaph as a way of making a judgment on the crimes of her betrayer.

30. *Virgil I: Aeneid i-vi.* Translated by H. R. Fairclough (Loeb 1916, rev. ed. 1978).

31. Cf. Grimal (17–18): "Pour une fois, Vénus et Junon collaborent; toutes deux poussent Enée et Didon l'un vers l'autre. . . . Et qu'est-ce que la mort d'une femme lorsqu'il s'agit de fonder Rome?"

32. Virgil's *Fama* is "monstrum horrendum, ingens" whose body is filled with as many ears, eyes and tongues as it is feathers (IV. vv. 181–3).

33. Reference to *Enéas* vv. 1539ff. / *Aeneid* IV. vv. 173ff. The *Enéas* poet says of *fame* "d'asez petit maint conte fait, / toz tens l'acroist, que qu'ele vait; / d'un po de voir dit tant mençonge / qu'il resanble que ce soit songe, / et tant lo vait muntepliant, / n'i a de voir ne tant ne quant" (vv. 1557–60).

34. Given the multivalent resonance attached to Venus, to love, and to Dido, these distinctions inevitably remain somewhat blurred and it can be difficult to assert any

single interpretation. Even with reference to Virgil, for example, Grimal asks how it is that Æneas ("fils de Vénus!") can logically be understood to subordinate his passion for Dido to another calling. Grimal points out, however, that Vénus is a formidable agent of political coercion and that "un caprice de Vénus ne saurait être seulement un caprice de femme. Parce que Vénus est déesse, toutes ses actions se situent sur le plan de la Providence . . . tandis que le caprice amoureux d'une mortelle est aveugle et s'abandonne au hasard" (17). I hope to show that the very logic that sanctions the historiographical myth (and its potential ironies) depends however on narrative continuity and not prescriptive moral dichotomies. At the same time, a revealing index to twelfth-century attitudes toward amorous myths at the center of the Greco-Roman Troy story which are consistent with the views expressed here can be found in Baudri de Bourgueil's lyric epistles between Paris and Helen, *Les Œuvres poétiques* . . . ed. Ph. Abrahams (Paris: H. Champion, 1926). I am grateful to Professor Ralph Hexter for alerting me to Baudri's epistles. Baudri's Helen, for example, dwells at some length on the divergent laws and ethical horizons binding mortals and gods. She recognizes that the divine will is superior, but that it is the fate of certain mortals to suffer from bad rumors and the silence of the gods. Paris's letter reveals the extent to which Baudri viewed Helen's "abduction" in the context of divinely ordained transfers of political and cultural dominion.

35. For the following outline I have relied upon the thorough and scholarly introduction to the Dictys and Dares chronicles provided by F. M. Frazer (Frazer 3–15).

36. In chapter three we will look more closely at Benoit's larger narrative and consider the emblematic significance of the emerging opposition between a "Greek" and "Trojan" ethos while probing further the significance of Paris's abduction of Helen as well as Benoit's treatment of this crucial event. We will also consider the nature of Benoit's interest in the homecoming (given that these events are subsequent to the fall of Troy and the exile from Troy of heroes appropriated or invented for the purposes of later medieval genealogical and historical fictions).

37. In Dares's account, the Jason and Medea episode has only a very oblique relation to the chain of causation leading up to the central conflict (Lumiansky 411): "In Dares' account there is absolutely no mention of Medea or of her love affair with Jason. Two brief sections, totaling only forty-two lines, are devoted to Jason's journey for the Golden Fleece. Clearly, Dares includes this material solely to motivate his account of the destruction of Old Troy, which results from Laomedon's inhospitable reception of Jason on the way to Colchus." For a discussion of Ovidian motifs, see Faral. Feimer provides a brief discussion of Celtic motifs appearing in the Jason-Medea episode.

38. There is also Medea's lyrical epistle (*Heroides* XII).

39. In Ovid's version, Medea is conscious of the war between reason and love that rages inside of her, and she acknowledges the danger posed by her desires (*Metam.* VII. vv. 9–21; 69–71; 92–3).

40. Benoit takes on a rather active authorial stance in the work's prologue through numerous and oftentimes strategically placed first-person commentaries (something we will examine in chapter three).

41. On the night of an arranged, clandestine meeting with Jason, her governess must warn her to act as though she were already in bed before sending for Jason lest she make a bad impression ("De la nuit est alé partie/ Sil tiendreit tost a vilenie/ Qu'a couchier fussiez a tel hore . . ."). Ovid's Medea (*Metam.* VII) invites

sympathy at the outset as well, at least insofar as she is proves sensitive to the nobility and beauty of Jason. In *Heroides* XII, by contrast, Medea is a less sympathetic figure. In this context, Ovid seems bent on making her one of the many abandoned brides whose vengeful ire is situated by a larger portrayal and critique of provincial rusticity and *saevae mores*. One can argue that this aspect ultimately prevails in *Metam.* VII as well. This Ovidian subtext, i. e. the contrast between rustic and urbane mores, is muted for the moment in Benoit's version, but it subtends the larger link between marriage fictions and their central role in relation to a larger western *translatio*, a *translatio* adumbrated in both the *Troie* and *Enéas* (more on this in chapter 3).

42. Feimer provides observations on Celtic elements in this episode. We will discuss the significance of similar motifs with regard to the final episodes of the *Enéas*. See discussion and note 82 below.

43. Medea's earlier demands:

Mais se de ço seüre fusse
Que jo t'amor aveir poüsse,
Qu'a femme espose me preisses,
Si que ja mais ne me guerpisses,
Quant en ta terre retornasses,
Qu'en cest païs ne me laissasses,
E me portasses leial fei,
Engin prendreie e bon conrei
Com ceste chose [the winning of the fleece] parfereies,
Que mort ne mahaing n'i prendreies. (vv. 1407–16)

44. Thus I disagree with Jones's more indulgent view of Jason: "Jason may be blamed for cowardliness in the moral sense, but hardly for misrepresentation" (44–5).

45. Rosemary Jones has defined this relationship as being "pre-marital" and implies that it fails accordingly. We will return to Jones's views of love and marriage, to the theme of marital consent in our further analysis of the *Roman de Troie* in chapt. 3 where we will likewise consider what doctrinal tenets may yet still remain unsatisfied. In any event, Medea cannot easily be said to be acting lasciviously, foolishly or recklessly.

46. Parallels between Dido and Medea are embedded in the tradition itself (Spence 30) (cf. *Heroides*).

47. I mean from the reader's standpoint, of course. (One can only speculate as to how a fictional "public" or "*vox populi*" would judge her fate.) In Ovid's text (*Metam.* VII), this eventual public condemnation, when anticipated by Medea, is associated with the harsh mores of her rustic fatherland.

48. So Virgil's Dido (I. v. 299; IV. vv. 651–4)

49. Thus while I agree that its overtly secular and historiograpical orientation diverges from the moral wisdom of much Latin commentary tradition, I would not state the division in terms as absolute as Poirion has used ("C'est donc en toute indépendance, par rapport à la *sapientia* des commentateurs en latin, que notre auteur élabore son esthétique et sa sagesse" 1985 vii).

50. Spence argues that at this level of typology, as opposed to that of moral *utilitas*, it is possible to locate a dynamic notion of *caritas* within medieval rhetorical paradigms: ostensible meaning and prescribed values give way to more mystical and experiential truths forming the basis of faith. Accordingly, the typological dynamics of Christian rhetorical models surrender a certain amount of control over fixed

meaning, over rigid assessments of the value of any given sign or temporal event.
51. Cormier sees Eneas's descent to Hades as tantamount (metaphorically) to an operation of divine grace which opens the way to the features of willful intent and consent to be developed as moral themes through the amorous feelings for Lavine.
52. Poirion (1976 218): "Dans la lecture moralisante et idéologique du texte une déchirure est apparue avec l'action de Vénus . . . ." Patterson sees the work to be flawed in a similar way. He sees the poet's attempt at moral legitimization as depriving the past of the very "historicity . . . that endows it with legitimizing value in the first place" (172–3).
53. This underlying contradiction reveals itself in Baswell's broader attempt to define the *Enéas* within larger and competing currents of medieval virgilianism. For example, the historical, imperial and genealogical themes that come to the fore in the *Enéas* do find their proper place within the peculiar exegetical tradition that orients the *Enéas* (as opposed to competing traditions, see note 52 below). This tradition ("the romance vision of Virgil"), Baswell points out, "is inextricably connected to the popular and vernacular histories produced from the twelfth century and through the Renaissance. Some versions are independent works, like the *Roman d'Enéas*; but this text is often found in manuscripts in the company of other *romans antiques* or the *Brut*, which implicitly renders it part of a continuing Anglo-Norman history. . . . The *Enéas* contributed to the prose *Histoire ancienne jusqu'à César*, which in turn influenced a later independent work, the *Livre des Eneydes*" (11). Abruptly, however, Baswell seems to place the *Enéas* within an entirely different sphere of interests—interests for which he provides no literary historical context: [taking up right where we left off] " The romance *Aeneid*, far more than its Latin source, is the story of Æneas and his women, or even the story of Æneas' women to the exclusion of Æneas. As I will argue later, it can be seen as the untold Latin *Aeneid*: a completion, but also a subversion of Virgil's narrative, tending to extend those very episodes, especially that of Dido, which for Virgil are the restraints keeping Æneas from his fortune in Italy" (11). It would seem, too, therefore, that the *Enéas* itself does not easily find its place within either of these typologies. For while Baswell emphasizes its place as an epic of imperial dominion written according to the realities of Norman conquest in England, he also is forced to acknowledge, as we stated, what he calls a constant process of "dilation" stemming from amorous motifs. Yet how is this later preoccupation logically accounted for by the historical tradition of exegesis and its particular propagandizing ambitions?
54. We will soon discuss the function of these speeches with regard to the work's closing marriage plot.
55. Baswell himself makes clear distinctions among a variety of distinct but never wholly exclusive exegetic "visions" in the high and later Middle Ages, all of which oriented receptions of Virgilian and Ovidian material. "Three major trends of interpretation can be distinguished. They merge, divide and recross throughout the period, but their identities are sufficiently clear to be useful. I will be calling these streams of interpretation the allegorical, romance, and pedagogical visions of the *Aeneid*" (9; cf. 9–13; 129; 136–8). In this regard it is important to recognize that despite the secular themes which dominate the *Enéas*, any *twelfth-century* adaptation of the *Aeneid* was at least as likely if not more likely to reflect the fluid fabulations of *cosmological* allegory as opposed to the exemplary didacticism of *moral* allegory. Indeed, Baswell points out repeatedly that use of Virgil for moral didacticism and the exploitation of the Trojan "counter-tradition" for the purposes of moral satire tend to reflect later developments, concentrated in the thirteenth and

fourteenth centuries ("The exegesis, literal and allegorical, took on a more strongly moral tone, and explication of the literal level shifted much of its attention from
historical difference to contemporary analogy, that is, to classical characters as
exemplary or cautionary figures" (138), while Christian, platonizing, cosmograph-
ical exegesis is a hallmark of the twelfth century. Furthermore, the moralizing tradition tends to divide and fragment the Virgilian narrative either to isolate passages
to be introduced into *florilegia*, to serve as passing examples in scriptural commentaries and sermons, or to generate ironic contrasts (137–8; cf. 129; 35–6; 11–13).
56. As it very clearly is in Benoit's *Roman de Troie*, for example.
57. For an attempt to read dense theological speculation *into* the amorous physiology of Dido, Lavine and Eneas see Laurie. For a complementary but less forced
reading, see Cormier's thorough exposition.
58. On the influence of narrative esthetics in the composition of historical narratives, see Partner (194ff.). Spiegel points out that the very challenges posed by a
"counter-tradition" could provide an important impulse to creative historiography
(46: "historical criticism, as we understand it, arose primarily when sources or testimony used by the chronicler disagreed on any given point, demanding resolution
. . ."). Spiegel emphasizes the importance of two predominant structural "grids" for
twelfth-century historiography a) typological exegesis and b) the structure of family lineage and genealogy (Spiegel 46–7). I am arguing here for the operative fusion
of both of these grids (cf. Spiegel 52). Spiegel also argues that these grids created
logical associations that allowed readers to discern logical relations not supplied
syntactically (52). In the *Enéas*, the poet emphasizes the genealogical trope (especially in Anchises's speech to Eneas in Hades) while simultaneously exploiting a
typological and syncretic syntax as well: in preserving Virgil's use of an artificial
chronological order and in allusions to Paris and the larger sphere of historical
antecedents contingent to Virgil's own narrative. The concepts of Fate and Fortune
which will be introduced into this discussion below also provided a structuring
topos for more philosophical and imaginative glosses of history. For the importance
of Fortune as a topos structuring twelfth-century (pseudo-) historiography see
Partner.
59. Spiegel (51): "Insofar as vernacular chroniclers remain faithful to the human,
biological significance of their genealogies, they can perceive relationships between
historical figures and events in the past as part of one continuous interrelated
stream of history . . . ." Thus historical events "stand in a filiative relation to one
another that mirrors the reproductive course of human life." Baswell points out
that "interest in lineage, and the particular attention to a theme of lineal return
through Teucer and Dardanus, is very widespread in the commentaries and recurs
regularly in the *Enéas*" (177). Benoit de Sainte-Maure would no doubt have to
overcome even greater obstacles to preserve the main elements of the problematic
Troy legend while attempting to legitimize the lineage of Troy for his probable
Norman patrons and readers. For a provocative discussion of the problematic
aspects of the larger Trojan myth in the Middle Ages, see Birns.
60. Eccles. 10; 8: "Regnum a gente in gentem transfertur propter injustitias et
injurias et contumelias et diversos dolos."
61. In this regard, Patch also finds traces of a continuum between Dictys and
Boethius (101).
62. In some cases the two figures are conflated; in others, they are dynamically
opposed to one another. The influence of Fortune may explain the separation of

lovers initially joined together by Love (Patch 95–6).

63. E. g., Jodogne (whose views we will discuss presently), Yunk (10ff. and 38).

64. Baswell informs us that this view of Æneas is emphasized in the commentary of Servius and perpetuated in the commentaries of Anselm of Laon who "notes, with Servius, Virgil's care in emphasizing that Æneas came to Italy 'driven by fate' (*fato profugus*), not through any criminal intent or desire for empire" (65).

65. See Ehrhart's discussion of the poet's version in relation to extant tradition, pp. 36–8.

66. Patterson (172–3) sees the poet using this historical inlay or cameo appearance of Paris as a way of creating a kind of meta-text of timeless moral analogy. This assumes, however, that Eneas is tainted by the same vice that marks Paris and serves to trigger the Trojan war and defeat. Therefore, it seems to be in contradiction with Patterson's observation that the judgment fable actually serves to siphon off the (negative) ambivalence surrounding Eneas by exculpating Eneas from responsibility for the Trojan War. As such, Patterson deems it a gratuitous addition devised to glorify Eneas. Nolan (163) sees an "implicit" moral prescription in the very allusion to the judgment, but I find it interesting that the poet largely goes out of his way to defer such a conclusion.

67. To underline this connection, we should note that the *Enéas* poet places particular emphasis on Turnus's legitimate claim on Lavine and on the imminent consummation of their already established nuptial agreement (we will return to this point below). It is also worth noting that this historical pattern predetermines Dido's undoing: the temple contiguous to her invincible fortress is dedicated to Juno: thus Dido's fate precedes her actions and the desire that leads to her undoing reflects more on Venus' superior merit than it does a moral allegory centering on Dido ("La deesse Juno voloit / que Cartage fust chiés del mont / et le realme qui i sont / a li fussent trestuit aclin, / mais onques n'i pot metre fin; / tot autrement est destiné, / car li deu orent esgardé / que a Rome l'estovoit estre," vv. 520–7). Accordingly, I have reservations about Baswell's contention that Dido is a central figure for dramatizing a disruptive power of feminine desire and excess that retards historical progress. For this same power would need be said to reside in the personification of *Discord* whose appearance is associated with the competing contentions of the three goddesses (Pallas, Juno and Venus). But if the goddesses figure a power antithetical to historical progress (as Baswell's logic necessarily implies), how do we envision Venus's role as history's primary *mover* in this text?

68. Virgil's Amata, not referred to by name in the *Enéas*.

69. Enéas's future is revealed in the predictions made by Anchises in the underworld and in the divine decrees referred to by King Latinus.

70. In Anchises's speech Lavine is entirely subordinated to her role as the mother of Silvius who marks the first in a long line of sovereign rulers. At the end, Anchises also announces all of the battles and travails that still await Eneas. Here Eneas's conquest is envisioned only in its martial aspects. At the same time, the possibility of any ethical or spiritual suspension or evolution seems to be rendered irrelevant by the fact that Eneas's success is already ratified ("molt se fait liez de sa ligniee, / qu'il voit qui tant ert esçauciee / que li monz vers lui aclin; / il regnera toz tens senz fin," vv. 2987–90).

71. "Through the nexus of commentaries the two texts virtually interpenetrate. In certain commentaries, especially those we associate loosely with the school of Chartres, there arises an allegorical monomyth, a single structure of explication

deemed adequate to both texts . . ." (Baswell 121–2).

72. In Bernardus's commentary on the *Aeneid*, Venus is given a two-fold association, first with cupiditas and second with divine love "Venus . . . est aliquando carnis concupiscentiam, aliquando mundi concordiam." However, Venus is defined as the latter when marked according to her relationship with Æneas: "Ubi ergo invenies Venerem uxorem Vulcani matrem Ioci et Cupidinis, intellige carnis voluptatem . . . Ubi vero leges Venerem et Anchisem Eneam filium habere, intellige per Venerem mundanam musicam, per Eneam humanum spiritum." At this juncture, Bernardus's commentary emphasizes Æneas's allegorical typology in terms of microcosmic agency itself: "Dicitur autem Eneas quasi ennos demas, id est habitator corporis, ennos Grece habitator Latine" (*Commentum* 10 ll. 4–9).

73. Obviously the influences I am arguing for retain only an oblique relation to the secular interests of the poet. It is interesting to note, however, that the poet's Pythagorean and platonic depictions of re-incarnation (the underworld episode) conform to those tenets of medieval neo-platonism that were recognized as being egregiously unorthodox (Duhem 70–1). We can only speculate as to Chartrian influences overall; however, Faral in a review of Constans's edition of the *Roman de Troie* [*Romania* 42 (1913): 88–106] believes that alongside Constans's suggestion that Orléans (a prominent site for classical studies) provides the most likely milieu for the cultivation of the *romans d'antiquité*, both Beauvais and Chartres represent equally credible possibilities (100).

74. Stoic views of love in classical Rome also converge with Platonic concepts (Grimal 271–2).

75. Gunnar Carlsson, "The Hero and Fate in Virgil's *Aeneid*," *Eranos* 43 (1945): 111–35, cited by Cormier (147).

76. Slightly later uses of Ovidian discourse for the purposes of exploring sentimentality do not seem clearly to motivate their appearance in the *Enéas*. "The *Metamorphoses* move on a higher style level than the love poetry of Ovid's youth; they approach the epic. But the author of the *Enéas* worked the love casuistry of the courtesan books into the epic material . . . He could not have used it in any other way . . . . And so the author of the *Enéas* transposed Ovid's love casuistry into another social class and another style, in which it seems—at least to me— rather out of place" (Auerbach 215). Similar observations are expressed by Yunck and Hanning.

77. My reading is supported by Poirion's contention that the poet's judicious handling of Virgil's matter does not always reflect a condescending posture toward pagan lore and divinities. On the contrary, "s'il y a une dépréciation des divinités invoquées par d'autres qu'Eneas, c'est au bénéfice de celles que le héros vénère. Le panthéon mythologique n'est donc pas réduit à une sorte d'évhémérisme ni à une figuration christologique. Il est recentré pour favoriser la religion d'un peuple s'opposant aux dieux de ses ennemis. Ce qui n'est pas sans analogie avec le rôle littéraire de la religion dans la vision militante des chansons de geste et des chansons de croisade (215). (Cf. my discussion below, however, concerning the *Enéas* and its critical departure from the essential tenets of the Old French *chanson de geste*.)

78. Crane (7 and 18–19); Warren (278 and 364–7). Most recently, Baswell (208–10).

79. In this speech, as Trojan suppliant is described in the image of Greek persecutor, we are forced to appreciate the potential political ironies at work despite what Patterson argued was the "unproblematic nature" of a work that on the surface

appears "committed . . . to its legitimizing function" (180 and 173).

80. Baswell remarks on the arbitrary nature of divine intervention within the social realism of the larger context but offers no explanation of how such a fictional device is to be interpreted: "Turnus' objection thus is to the breaking of specific feudal covenants; Latinus and Eneas will respond with the superior (and, in the twelfth-century, contested) claim to lineage, *and the desires of a pantheon to which they have exclusive access*" (208 emphasis added).

81. The ironic possibilities afforded the poet, however, are anticipated in Boethius's own critique of the "genealogical fallacy" which provides the obverse side of the *vox populi* topos: "Plures enim magnum saepe nomen falsis uulgi opinionibus abstulerunt; quo quid turpius excogitari potest? Nam qui falso praedicantur, suis ipsi necesse est laudibus erubescant" *Consolatio* III. pr. vi. 5–8; on noble lineage and empty praise ll. 20–29).

82. Poirion (1985) sees the depiction of the Judgment of Paris in the *Enéas* as having epic and mythical more than moral reverberations (in accord with our comments above). Specifically, Poirion sees in Paris's choice a reflection of resurgent Indo-European categories of thought: "Le choix est donc résumé par les trois termes: richesse, prouesse et femme, trois dons dont le système nous fait penser à la tradition des contes folkloriques, mais que l'on peut rapprocher des fameuses trois fonctions indo-européennes qui justement font résurgence dans les écrits de cette époque: la souveraineté, la force militaire et la fécondité" (ix).

83. Poirion (1976 216) has observed the same phenomenon in other passages (vv. 6689–90 and 7866–7).

84. VV. 3489–90: an ironic echo of Dido's own appropriation of land and dominion in Carthage (vv. 377–406)?

85. Cormier (189, note 53): "the slow-paced description of the wound and the stag's suffering is quite naturalistic and recalls *la chasse au cerf blanc* in Chrétien's *Erec* . . . and in Marie de France's *Guigemar* . . . ." In a note to his translation relating to the episode in question, Yunck points out that Hoeppffner (*Arch. romanicum* xv, 263–4) observes parallels between the early part of Ascanius's hunt (vv. 3585–3600) and the description of Brutus's hunt in Wace (*Brut* vv. 136–46).

86. Bromwich also points out that during the twelfth and thirteenth centuries the White Stag motif provided a "literary commonplace" which could "be attached to any character indiscriminately, and . . . form a prelude to almost any kind of magical adventure" ("Dynastic Themes" 444). Bromwich specifies, however, that the Stag Hunt was most intimately connected to the theme of sovereignty and dominion in the Breton tradition through which it was received: "This tale is associated with the founder of one of the Breton kingdoms, and also with a character whose name was borne by several early rulers of another" (463).

87. In Virgil, the wounded stag is soon forgotten as skirmishes erupt between the pastoral Rutulians and the Trojan host. In the medieval account, however, the stag is seized, slain and dressed for consumption.

88. A summary of this tale is produced (in modern English) by Bromwich ("Dynastic Themes" 447).

89. For a divergent interpretation of the stag hunt episode in the *Enéas* see Adler ("Eneas and Lavine" 83: "Not without arrogance and almost cynically,—the worst indictment against aristocratic *superbia*!—Ascanius, at a time when there is already bloodshed, insists that the stag be prepared for dinner . . . . Such a display is not likely to defeat *old* fashioned feudalism. It is too much like it"). It is unfortunate

that Adler failed to consider the Celtic analogues, because they would seem to com-plement nicely his view of Lavine as a reflex of the *puella senex*. For an overview of the *puella senex* topos, see Curtius (101–5). Curtius's discussion reveals that the *puella senex* figure possessed features analogous to the Celtic sovereignty Bride-Hag motif (also discussed by Bromwich). The Bride-Hag figure was often integrat-ed into the Stag Hunt topos (Bromwich). For a discussion of more general "Celtic" features that appear in Benoit's account of the Judgment of Paris, see Ehrhart (40–3). For Celtic elements in Benoit's version of the Medea story, see Feimer (46–7 and 49, note 7).

90. Penny Eley (28–9): "The myth of Trojan descent is first attested in the seventh century for the Franks, in the tenth century for the British, and the early eleventh century for the Normans. In the twelfth century the myth acquired an important political dimension in Britain: one nineteenth-century critic claimed that it had the status of a 'croyance d'Etat' at the court of Henry II, who tried to legitimize his rule by promoting the view that Normans and British were of one blood and should therefore share allegiance to one leader."

CHAPTER THREE

Twelfth-Century Marriage Reforms and the Representation of Marriage

Our extended study of the *Enéas* goes beyond a preoccupation with reductive moral prescriptions to reveal the synthetic typologies influencing the meaning and representation of love and marriage in relation to larger epic themes and narrative structures. In focusing on the *Enéas*, we have also gained an important diachronic perspective on twelfth-century literature. We have seen that the expression of amorous sentimentality is not yet realized to its full potential nor integrated with convincing subtlety to the essential outline of Virgil's plot. Nonetheless, the epic contours of the Latin legend allowed for rich and imaginative explorations of amorous and marital themes, and the typological vision of Christian thinkers (in its historical and moral dimensions) resulted in imaginative intellectual responses to classical traditions.

We still need to consider why it is that marriage themes came to be so prominent in this process. We have already seen that classical and Celtic traditions provided the raw material for a rich synthesis of motifs and themes from epic and lyric genres and for linking amorous and historical elements to a divine concept of *eros*. This allowed medieval authors to envision a new form of secular narrative involving the rhetoric of amorous sentimentality and retaining conflicts inspired by older epic traditions in which the fatalistic aspects of Indo-European sovereignty myths (sovereignty-bride and bride-quest) highlighted the sacred dimensions of the conjugal union. While this confluence of pagan motifs was (in conjunction with dynastic interests motivating a revival of the Troy legend) galvanized in the marriage-abduction plot at the center of the Trojan war and inscribed in Rome's foundation, the harmonization of pagan and Christian typologies was facilitated by contemporary interests in marriage doctrine reform and the sacramental value of marriage as a formative institution in the secular orders of Christian society. This twin observation leads to a number of important considerations with regard to later developments in

Old French literature: 1) The revival of Indo-European traditions—although naturally not perceived as such—provided the basis for a rich exploration of marriage themes extending beyond any narrow fascination with questions of marital propriety and sexual (im)morality. The historical, sacred and providential aspects of Indo-European marriage ideology would dovetail neatly with contemporary theological speculations intent on eluci-dating the sacramental significance of the conjugal bond. 2) Since the impe-tus for "glossing" Indo-European marriage and sovereignty motifs is large-ly circumscribed by emerging theological discourse on marriage, the prin-ciples of marital legitimacy adopted from Roman Law—namely the con-cepts of spousal consent and affection—would play a major role in fiction-al representations of marriage and provide a framework for the integration of epic and lyric motifs. 3) The development of Christian marriage reforms includes not only important judicial developments but a larger theological and cultural field of speculation which more or less explicitly exceeds the boundaries of patristic dogma to find inspiration and precedents in a deep-er vein of enduring European tradition. This search for a consistent amorous and marital mythology led to inherent cultural paradoxes. Contemporary marriage practices—the object of reform—were ultimately rooted in the same Indo-European traditions that lent mythical dimensions to the exclusivity of the conjugal relationship. So "reform" involved not only a departure from the corporate prerogatives of the aristocratic family but a return to heroic archetypes just as the notions of consent and affec-tion potentially conflicted with the corporate and conservative aspects of Church morality.

The apocalyptic Trojan-Greek confrontation provided a paradigmatic historical and typological topos for exploring such paradoxes and tensions. The redemption of Troy and the Trojan hero Paris in Old French secular narrative signals a radical departure imagined as the return to an ancient, latent archetype. The opposition between Greek and Trojan as well as the ultimate fortune of the Trojans over the long course of *translatio imperii* in the West adumbrates a tension between alternative visions of feudal order and dominion that are dramatized in terms of conflicting secular (aristo-cratic) mores—martial, amorous ("erotic"), and marital. The preoccupa-tion with the ancient figure of Paris as an amorous archetype involves a contemporary "gloss" in which the lyrical models of Ovid and the trouba-dours give broad sentimental dimensions to the vague term of "affection" employed in Roman law. But the tragic aspects of the Trojan legend, the cataclysmic conflict which precedes Rome's ascent to Western dominion, remained as a reminder of the anarchic powers of passion, of the need to envision a harmonious correspondence among amorous desire (as a mani-festation of divine will), marital legitimacy and social stability. Hence medieval interests in the subjective experience of love were made com-pelling for Christian authors not only in response to the secular pastimes of aristocratic patrons, but by virtue of the essential link between love and legitimate marriage as it came to be defined in the context of a larger providential and sacramental vision of marriage governing temporal affairs.

Of course, these developments, convergences and syntheses are in many cases only dimly realized in the *Enéas* itself. Indeed, the divergent critical assessments of the work corroborate what our intuitions tell us, that the author never fully resolves the problematic nature of Eneas's Trojan heritage, nor fully explores either the theme of marital legitimacy itself, or the role of love and consent in relation to marital fortunes. In the *Enéas*, the divine nature of the Trojan supremacy seems to be almost entirely circumscribed by the inscrutable logic of a prophetic, genealogical destiny, albeit one in which martial trials are to some extent awkwardly eclipsed by venereal ones. Alternately, in Benoit's *Roman de Troie* the underlying paradoxes inscribed in the very nature of Eneas' Trojan origins are engaged head on. Perhaps Benoit consciously reflected on the problematic features of Eneas's own itinerary and on its ramifications for the attempt to promote Norman aspirations by reviving and appropriating Trojan "history" and "ancestry." Of course, we cannot know what Benoit may have actually thought, but his telling of the Troy story is clearly designed to redefine central features of the Trojan myth in terms of contemporary constructs. In chapter two, for example, we found that Benoit's version of the Medea story gave emphasis and clarity to the attitudes toward marriage embedded in the *Enéas* poet's retelling of Virgil's epic. Likewise, the larger narrative not only offers a comprehensive exploration of the convergences outlined above (between epic and mythological traditions, on the one hand, and emerging marriage doctrines, on the other), it also makes marriage practice into a pervasive, unifying motif.

In approaching this vast work (surpassing 30,000 lines), our main concerns will be 1) to see how the dubious nature of Eneas's ancestral lineage is reoriented by a rather radical act of historical or mythical revisionism with regard to the figure of Paris and the act of abduction that triggers the Fall of Troy; 2) to highlight briefly the prominent role of marriage as a plot motif and, more importantly, as a central thematic nexus unifying Benoit's expansive narrative, and 3) to identify the nature and importance of concepts or attitudes directly influenced by emerging marriage doctrine reforms.

In elucidating these features of Benoit's text, I will set the stage for a broader investigation into the role of emerging concepts of marital legitimacy in a variety of later twelfth- and early thirteenth-century Old French secular narratives. Our study of Benoit's text will broaden the conclusions made in reference to the *Enéas* by highlighting the formative influence of emerging marriage doctrine with regard to vernacular adaptations of and twelfth-century responses to pagan traditions. Next, we will look briefly at some key aspects of developing twelfth-century marriage doctrine so as to provide a context for a deeper appreciation of the relation between Old French marriage fictions and emerging theological attitudes.

LE *ROMAN DE TROIE*

Preferring to focus on the amorous rhetoric and episodes of unrequited passion, scholars have tended to ignore marital motifs in the *Roman de Troie*. Interest in Benoit's novel treatment of amorous tropes and amorous figures (e.g. Medea, Briseida, Troilus, and Achilles) eclipses Benoit's historical vision and the functional role therein of the marriage motif itself. We saw in the *Enéas*, for example, that the lyrical interlude between Lavine and Eneas is deferred and somewhat awkwardly integrated into the hero's providentially ordained itinerary. While amorous motivations and subplots are perhaps more thoroughly and realistically tied to narrative events and causation in the *Roman de Troie*, Lumiansky's contention that Benoit is fascinated primarily with amorous passion leaves one wondering what interest Benoit found in the larger epic dimensions of the Troy story. Lumiansky argues that love motifs play an *integral* role in the *Roman de Troie* and that "about sixty-five hundred lines in the poem are directly concerned with the love stories of Jason-Medea, Paris-Helen, Troilus-Briseida, and Achilles-Polyxena" (424). However, critics have failed to define the larger vision governing the work. Alfred Adler focuses on the fusion of *militia*, *amor*, and *amicitia*, but even this inclusive approach fails to account for many events connected with the Greek homecoming:

> At this point [after Pyrrhus slays Polyxena (his father Achilles's beloved)], the account of events after the fall of Troy becomes rather sketchy, and often is confusing . . . We have the impression, if Constans's text is reliable, that Benoit wearily hastened toward his *Epilogue*. ("*Militia et Amor*" 27)

Adler suggests that the theme of *amicitia* surfaces only at the end, as a "whisper" and that the sudden apparition of this new historical *prise de conscience* is neither anticipated nor coherently linked to the "homecoming"(15 and 28).[1]

It would indeed seem likely that Benoit faced a dilemma as he sought to gloss the circular and tragic events outlined in the Dares-Dictys chronicles. In focusing on amorous tropes, however, one fails to recognize the blatant ubiquity of the marriage theme and the unifying role it serves. As in the *Enéas*, a marital drama (the abduction of Helen) triggers the principal historic event and provides a preliminary link in the chain of historical causation that ultimately leads to the foundation of Rome. In her chapter on love in the *Roman de Troie*, R. Jones largely limits her analysis to the four major love affairs in the first half of the narrative (i.e. preceding the events of the

Greek homecoming), but she nonetheless stresses the important role that Benoit attributes to marriage:

> It is interesting that the three relationships which really end unhappily are extra- or pre-marital; in all three [i.e. 1) Jason-Medea 2) Troilus-Briseida 3) Achillès-Polyxena] loyalties are broken and society is flouted. In the relationship between Helen and Paris a lasting relationship similar to marriage, with the responsibilities it entails, seems to have value as a bulwark of society, and the importance of society and of loyalty appear to be the essential criteria for Benoît. (59)

While I agree with Jones's insight here, it is not clear why she says that Helen and Paris are joined in a relationship that is only "similar to marriage."[2] Nor is it clear how or why the adulterous Paris-Helen situation is 'right' in every sense as Jones proposes, or why Benoit uses an archetypal tale of social and marital transgression to convey his ideals of social order and loyalty. The failure to define a consistent "ethos" governing marital fictions in Benoit's text reveals the need for new approaches.

Comparing the Ovidian and French versions of the Medea episode, Jones argues that Medea is placed in a more favorable light by Benoit, but then suggests the same for Jason. She argues that both lovers ultimately suffer because of their failure to fulfill certain precepts of "courtly love" that would have insured a lasting marital bond (44–45). When she touts the Paris-Helen relationship as representative of a positive ideal, however, courtly love conventions play no important role. Later she asserts that (based on the four primary love "situations") "courtly love does not represent an ideal for Benoît" (59). Regarding Paris and Helen, she thoroughly demonstrates the extent to which "Benoît has deliberately placed [Paris-Helen] in a favorable light" (47), but fails, as we have said, to resolve the tension between this positive portrayal and the underlying marital transgression with all of its disastrous consequences (47–50). The fact (pointed out by Jones) that Benoit does seem intent on letting the wronged Menelaus largely fade into oblivion suggests, nevertheless, that some rationale must underlie the happy union of Paris and Helen. In her closing statement, however, Jones shifts to a new perspective rather than accounting for her earlier observations:

> Yet in spite of this [i.e. the lasting relationship between Paris and Helen] the impression of tragedy remains with the reader, and one is left with the feeling that people are puppets swayed by love, which causes innocents to suffer and imposes conflicts and problems beyond solution. (50)

The shift presents an honest response to the earlier impasse, but Jones never goes on to account for the unpredictable consequences of love, nor does she try to resolve the contradiction that exists between the assumption that legitimate marriages are governed by the proper cultivation of love ("courtly love") and the alienating and overpowering aspects of an impersonal and universal god of *eros*.

The first explicit indication that the marriage theme is a fundamental focus of Benoit's "reading" and "gloss" of the Troy (hi)story comes in a fleeting but far from subtle authorial interjection at the conclusion of the first destruction of Troy. The author's remarks are in reference to Hercules' abduction of Hesione. Hesione is King Laomedon's daughter and a sister of Priam who is Laomedon's son and heir to the throne of Ilion. During a raid and sack of Laomedon's citadel at Troy, Hercules abducts Hesione and hands her over to Telamon as a war prize. Following an ample and somber description of the city's destruction, of the terror of its townspeople and of the ruthless slaughter and plunder carried out by the Greek war band, Benoit reports on Hesione's abduction (vv. 2761–2792). The narrative account of the event is punctuated by two brief but trenchant authorial remarks:

> La fille al rei, Esiona, —
> Ja mais plus bele ne naistra,
> Ne plus franche ne plus corteise,
> Grant ire en ai e mout m'en peise, —
> Cele en a Telamon menee:
> Dans Herculès li a donee,
> Por ço qu'en Troie entra premier.
> N'en ot mie mauvais loier,
> E s'il a femme l'esposast,
> Ja guaires donc ne m'en pesast;
> Mais puis la tint en soignantage
> Ço fu grant duel e grant damage. (vv. 2793–2804)

> [*The king's daughter Hesione—a more beautiful woman will never be, none more generous or more courtly. It gravely angers and troubles me that Telamon has abducted her. Hercules gave her to him as a prize for being the first to enter Troy. Never has there been a worse reward. If only he had made her his wife, hardly would this be so troubling. But afterwards he kept her as his concubine, which was very sad and very wrong.*]

Here the poet, although with some restraint, poses as exegete. The narrative voice becomes first judgmental (e.g. "N'en ot mie mauvais loier") and then explicitly first-person
("Ja guaires donc ne m'en pesast . . ."). The narrator speaks as both witness and judge. He expresses both an explicit affective response and an implicit moral condemnation ("Grant ire en ai e mout m'en peise . . . E s'il a femme l'esposast / Ja guaires donc ne m'en pesast"). It is more common for Benoit's narrator to intervene obliquely with the purpose of foreshadowing future events so as to add to the tragic fatalism of his narrative. The final line in the passage quoted above provides an example of this device (Ço fu grant duel e grant damage). This form of judgment is, in terms of attribution, doubly vague ("who" says it, and to "what" does it refer exactly?). But, by virtue of the veiled moral condemnation that precedes it, this rather vague lament is capable of a broad resonance. Its lack of clear

attribution and its foreboding undertones foreshadow a certain if uncircumscribed catastrophe. The anticipatory perspective adumbrates the elusive machinations of an inexorable and abstract Fate, while the *passé simple* (fu) establishes a decisive nexus between a single moral (mis)deed and the fixed course of an inalterable and tragic history. Telamon's abusive treatment of Hesione is not only the *subject* of "duel" and "dommage," it is also the (historical) *cause* of a tragic legacy. Both of these are lamentable events, and the editorial judgments cast a larger perspective on Telamon's immediate misdeed by linking it to future tragedies. In order to underscore the dynamic rapport between the private sphere of marriage (conjugal strife) and a vast historical plane of action, between microcosm and macrocosm, Benoit uses his incriminating commentary on Telamon's conduct toward Hesione as a conclusion to his larger account of the first fall of Troy. The first Fall of Troy and the major Trojan war stand in an analogous relationship of scale: the first is a small war of private vengeance which mirrors and foreshadows an impending war of eschatological dimensions.

As if to erase any lingering doubt about the underlying causal link between Telamon's conduct and the imminent escalation of war, Benoit reiterates a conventional topos that links vast historical events to "small causes":

> Or vient uevre, s'est qui la die,
> Ja mais teus ne sera oïe. . . .
> Saveir par com faite acheison
> Avint ceste destrucion.
> Par assez petit d'uevre mut,
> Mais mout par monta puis e crut . . . (vv. 2825–32)

> [*Now comes a tale of deeds. Tell it who may, never will anything like it be heard. Be it known what fateful events brought out this destruction. It all started with a rather small deed, but swiftly thereafter did the consequences multiply and increase.*]

The geometrical perspective neatly links Telamon's private transgression of conjugal propriety to a momentous historical tragedy. From this vantage point we can fully appreciate the gravity and larger thematic significance of the author's strategically placed commentary on the Telamon-Hesione episode. In essence, the event signals a historical "point of no return" where moral will and reflection might otherwise have prevented men from becoming blind conspirators to a cycle of tragic repetition. It is Telamon's refusal to recognize Hesione's royal status (she is a king's daughter) and to pay proper external tribute to that status (through the rites and duties of a marital contract) that marks this fundamental historical turning point. In this way, Benoit inextricably links the regulatory powers of marriage to the problems of ethnic warfare and to the rights and struggles of sovereign rulers.[3]

The marriage topos also provides a common thematic link between the events of the Trojan War and those of the Greek homecoming. The marriage of Agamemnon and Clytemnestra unites, in terms of both causality and theme, the bulk of narrative threads making up the *Nostos* as Benoit adapts it from Dictys and Dares. Indeed, it is interesting to note just how many narrative threads lead back to the domestic crimes surrounding Clytemnestra. Of course, murder and matricide—the crimes of revenge committed by Orestes—are among the primary repercussions. A minor but telling example of the pervasive role of marital dramas in the Latin chronicles themselves comes in the form of a secondary event tied to the Orestes story. In this brief passage (which is preserved by Benoit) Dictys reports on the personal motivations of a Greek figure named Strophius who assists Orestes in his plot to kill Aegisthus. Prior to his adulterous affair with Clytemnestra, Aegisthus was the protagonist in another marital transgression: he married Strophius' daughter and then insulted Strophius by repudiating her after he fell in love with Clytemnestra.[4]

In other contexts, the chronicles also show that marriage can be an institution that promotes harmony and reconciliation between different family clans. The enmity that arises between Orestes and Menelaus because of the death of Clytemnestra is resolved through a new marital agreement when Menelaus grants Hermione (his oldest daughter by Helen) to Orestes as a wife (vv. 28533–28548). Soon, however, this event leads to a renewed act of marital transgression recounted in a narrative sequence that seems to evolve from Benoit's own gloss of an inconsistency in Book VI of Dictys' chronicle. After mentioning that Menelaus grants Hermione to Orestes as a sign of reconciliation (VI. 4), Dictys—without reference to any intervening episode of abduction—abruptly remarks on his presence at the wedding of Hermione to Neoptolemus (VI. 10 and 12). From these references Benoit elaborates a tale of abduction no doubt inspired and corroborated by the Ovidian Hermione (*Heroides* VIII). Pyrrhus (a.k.a. Neoptolemus) *abducts* Hermione (Orestes' wife) and marries her (vv. 29596–29602). At the same time, however, Benoit reminds us that Pyrrhus is already living with Andromache who came to him as part of the spoils from Troy. Thus a new tale of marital strife evolves as Hermione realizes Andromache has actually become the object of her husband's love.[5] Ultimately, the tale of abduction unfolds into a new tale of marital revenge as Orestes kills Pyrrhus and retakes possession of Hermione. Benoit's reference to Pyrrhus' exceptional nobility (the son of Achilles, he is the greatest Greek hero of the day) only underscores the tragic overtones associated with his death and with the cycles of domestic vengeance that Benoit weaves into his tales of marital impropriety.

As we consider the many interwoven marital dramas in the chronicles of Dares and Dictys, we are forced to consider how natural it would be for the marriage question to have impressed itself upon Benoit as a crucial social institution and cultural barometer. The problematic affair between Clytemnestra and Aegisthus at the center of the homecoming provides a thematic counterpart to the notorious affair (between Paris and Helen) at the center of Troy's destruction. The fate of each nation, revolving as it does around a crucial tale of marital strife, comes to mirror the other.

Lumiansky has analyzed and commented upon the unifying role played by episodes of amorous courtship. It is worth noting, however, that in many cases, these "love" stories can equally, if not more accurately, be described as tales of marriage. Lumiansky's failure to see this ambivalence betrays the critical tendency to view medieval love and medieval marriage as virtually incompatible terms. Yet just as Benoit opens his tale with the Medea story elaborating a tale of marriage and (impending) desertion and goes on to reconstruct a minor episode—Telamon's abduction of Hesione—around the theme of conjugal impropriety, so too does he ceremoniously wed Helen to Paris, and frame Achilles's passionate love for Polyxena with a larger drama, namely the contract of betrothal (initiated and then renounced) between Achilles and the parents of his elected bride.

In Benoit's account, therefore, marriage becomes the essential motif which unifies and propels both the Trojan and Greek segments of the larger narrative and provides a natural framework for the exploitation of lyrical tropes in consonance with the epic dimensions of the historical legend. The introduction of Ovidian rhetoric and motifs functions as a kind of lyrical *amplificatio*. The suppression of these protracted lyrical interludes in the latter portion of the narrative allowed Benoit to increase the momentum of Fate's tragic trajectory and effectively enhance the pathos of the Greek tales. By mental reference to earlier lyrical digressions, however, the reader could continue to provide his own gloss of the affective forces underlying private marital conflicts. In his most original portrait—that of Briseida—Benoit goes well beyond the tropes of refined courtship and amorous intrigue to construct a complex *exemplum* with direct links to the larger marital and political themes. Benoit complicates our judgment of Briseida by presenting her under incongruous guises. As she prepares to leave Troy, Benoit emphasizes her beauty and the "marvelous" qualities of her accouterments (vv. 13329 ff.). These majestic and magical aspects find a logical corollary somehow in her extraordinary fate; I mean that she could easily be seen as a kind of allegorical figure whose own nobility and inconstancy reflect the transcendent but seemingly arbitrary and unpredictable whims of Fate and Fortune. Yet, in describing Briseida, Benoit shifts abruptly from his description of her marvelous features to a conventional critique of female inconstancy (vv. 13428 ff.). The contrast is hard to resolve. As in the case of the Old French Dido, we are ultimately invited to question the conventional "moral" wisdom applied to Briseida as a "type" exemplifying inconstancy. Benoit's strategy is also reminiscent of Virgil's. He undercuts the conventional misogynist censure by conflating it with an unreliable *vox populi*. The scorn cast upon Briseida is attributed to a crowd of anonymous Trojan women.[6]

Briseida's experience obliquely serves to enhance our understanding of Andromache to the extent that Andromache, like Briseida, also finds herself compelled to negotiate political boundaries. Andromache, in turn, can be said to complement Briseida's own amorous impasse: she moves on from the ruins of Troy to become, in Greece, the emblem and progenitor of a new, more inclusive, political dynasty founded on the common maternal birthright of Laudamanta (fathered by the deceased Hector and grandson of royal Priam) and Achillidès (fathered by Pyrrhus, and grandson of the great hero Achilles) (vv. 29777–804). The contemporary political resonance surrounding these contrasting and parallel female figures is adumbrated by a digression in which the narrator, addressing an unidentified noblewoman (who may well represent Eleanor of Aquitaine), distances himself from the statements of those condemning Briseida's inconstancy ("De cest, veir, criem g'estre blasmez / De cele . . ." vv. 13457 ff.). In this apostrophe to a supposed queen and patron, Benoit effectively brings about an evident *rapprochement* between the respective fates of Briseida and her counterpart Andromache, on the one hand, and a figure evoking Eleanor of Aquitaine, on the other. Both Briseida and Andromache are, as I have suggested, transnational (sovereignty-) bride figures. The fundamental difference seems to be that Briseida never regulates her amorous bigamy through the enduring bond of marriage. Andromache remarries and becomes the matriarchal founder of a new nation that reunites Greek and Trojan blood.[7] Ironically, therefore, the cycles of national, feudal and domestic carnage so emphatically dramatized by Benoit, and so deeply rooted in the tragic character of the "Greek" world, can be said to find their emblematic resolution in the form of an inter-ethnic marriage and family unit (mother and step-brothers) freed from the historical patterns of exclusion, war, and repetition which are perpetuated by paternal authority.[8]

The metaphoric relation between bride and sovereignty is thus integrated into a broader psychological vision of amorous behavior to account for the transfer of dynastic power through Western history. Love, like sovereignty, is governed by a transcendent force (a force akin to *eros* that simultaneously manifests itself as an agency of providence and as an alienating passion, as *Amor* and *Fatum*), and its prerogatives ultimately transcend feudal custom and patriotic interests. Thus it is fitting that Benoit should end his Troy narrative with a dramatic staging of inadvertent patricide that provides an exceptional instance of the sublime in twelfth-century literature. The Oedipal complexities surrounding Ulysses's death at the hands of his own bastard son (Telegonus) convey a tragic pathos that in an unexpected and uncanny fashion harks back to an early Greek spirit no longer evident in the Late Antique chronicles. Ulysses's fate (brought upon himself by a forgotten period of lust and fornication) does not, however, entirely efface the great hero's destiny which is to be preserved through the contrasting marital fortune of Telemachus: a legitimate son who is married to Nausicaa in a spirit of mutual consent and with the sanctions of a publicly celebrated ritual.

In total, Benoit's narrative contains references to at least thirteen distinct marriage dramas and to a wide range of matrimonial issues. There are four instances of repudiation, betrayal or desertion; two instances of semi-clandestine marriage; four cases of abduction (one of these explicitly involving rape); six cases of adultery or concubinage; at least two cases of medieval "bigamy" (Helen, Andromache); two instances (both involving Ulysses) of illicit fornication. Benoit also draws upon a broad range of amorous tropes and terminology adapted from various kinds of sources. We find rhetoric and motifs from Ovid and from Breton lore, references to *fin'amor*, and tropes reminiscent of troubadour lyric. Benoit makes reference to a vast typology of amorous and marital states: there are references to narcissistic love (Achilles), to wedlock devoid of love (Hermione-Telamon, Helen-Menelaus; Jason [-Medea]) and to *amour-passion* which culminates in a pledge of betrothal (Paris-Helen; Achilles-Polyxena) and to married lovers (Paris-Helen). There are oblique allusions to the issue of marital consent (parental in the case of Medea; spousal in the case of Helen-Paris and in the case of Hermione's abduction by Pyrrhus).[9] In commenting upon Ulysses' adventures with Calypso and Circe, Benoit expresses a moral distinction between sinful lust and illicit fornication on the one hand and the refined and cultivated passion of *fin'amor* on the other. He consistently condemns infidelity (both amorous and conjugal), but he also evinces a subtle appreciation of moral and political realities when he expresses sympathy for unconsenting spouses (Helen, Hesione) and for victims of political tribulations (Briseida, Andromache). He seems to portray marriage as a necessary means of regulating passion and judges love to be necessary to a lasting marriage.

Ultimately it is hard to tell whether Benoit's stance as a "moralist" precedes his historical reflection. It seems, rather, that his twelfth-century learning provided him tools for exploring and "glossing" history in broad social, national, and religious terms. Benoit also employs amorous motifs in order to elucidate the links between private action and epic history. These two planes of experience converge in the politics of legitimate and illegitimate marriage. In approaching marriage issues, however, Benoit goes beyond ready-made moral prescriptions. In his condemnation of Telamon, for example, the act of abduction is not in itself the object of moral censure. Specifically, Benoit laments Telamon's decision not to marry Hesione: Telamon "l'a tint en soignantage" (i.e., as his concubine). The abduction is justified by the equality of rank involved—Hesione is exceptionally noble (Ja mais plus bele ne naistra / Ne plus franche ne plus corteise")—and it is intimated that a marriage with Telamon, whose martial valor was clearly demonstrated in the siege on Laomedon's citadel, would not be a discredit to her rank ([Telamon] "N'en ot mie mauvais loier"). This inherent equilibrium is violated when Hesione is forced to remain a concubine. The degrading implications of Hesione's status as "booty," and the inter-ethnic tensions inherent in the abduction of a royal subject are both exasperated by Telamon's proud obstinacy. By the same token, therefore, we are encouraged to assume that marriage would have preserved Hesione's private honor while also bringing about an effective political *rapprochement* between potentially hostile families. Since Hesione is King Priam's sister, her marriage to the Greek Telamon would have established a powerful international alliance that would counteract the tenacious ethnic divisions permeating Benoit's vision of the ancient Greek world.

The significance of Telamon's transgression is evident in Benoit's depiction of the escalating antagonism that leads to the Trojan war. Despite all the wrongs suffered at the hands of the Greek invaders, Hesione's status quickly becomes the central focus of Priam's war counsel. At first, military confrontation is deferred in the interest of her safety (vv. 3233–3244). However, the perspective gradually begins to shift as Priam instructs his ambassador (Antenor) to try, above all, to secure Hesione's return. Ostensibly, Priam would seem to want to insure her safety before beginning a war of retaliation. His words, however, also betray a marked hostility with regard to Telamon's dishonorable treatment of his sister. Priam's charge closely echoes the narrator's own judgment:

> Ma soror ai en lor contree,
> Qu'uns vassaus a asoignantee,
> Ne la deigne prendre a moillier.
> Peser m'en deit e enuier.
> Trop est grant honte et lait damage
> Que fille a rei est en servage. (vv. 3219–26)

> [*My sister is in their country, where a vassal has taken her as his concubine. He cares not to take her for his wife. For this I am troubled and sad. It is a too shameful and criminal offense for a king's daughter to be treated as a slave.*]

Almost imperceptibly Hesione's release becomes *the* condition upon which war and peace will depend. Priam, instructing his ambassador, says:

> A ceus dites qui furent ça,
> S'il me rendent Esiona,
> Ja ne m'orront puis tenir conte
> Del lait, del tort ne de la honte
> Qu'il firent d'ocire mon pere,
> De mes sorors ne de ma mere.
> Li sole [i.e. Esiona] quier jo e demant:
> Honte ai qu'on la tient a soignant.

Metez i or norme et poëir
Coment nos la poissiens aveir. (vv. 3261–70)

[*To those who shall be present, please say this: If they return Hesione to me, they will no longer hear me speak of the crime, the wrong, and the shame they inflicted by killing my father, mother, and my sisters. It is only her [i.e., Hesione] that I demand. It is offensive to me that she be kept as a concubine. After saying this, set down there and then the terms and safeguards for securing her safe return.*]

Upon arriving in Greece, Antenor stops at the residence of Peleüs. He refers to the death of Laomedon and the sack of the city as shameful acts (vv. 3306–3309), but he dwells most emphatically on Hesione (vv. 3310–3326). In the end, Antenor's references to Hesione seem to reflect more than a diplomatic strategy, and we (like Peleüs) are left believing that the eventual outbreak of hostilities hinges entirely on her welfare:

En soignantage la tient cil
Qui l'en amena en eissil :
Grant honte en fait Priant le rei
Or si te mande e dit par mei
Que sa soror aveir li faces:
Il ne te fait autres manaces,
Mais qui en pais la li rendra,
En bone pais la recevra.
Il n'a talent d'el demander:
Ja ne l'orreiz de plus parler.
Vers vos eüst mout a requerre,
Mais ja par lui n'en sordra guerre,
Noise ne tençon ne meslee,
Se sa suer li est aquitee. (vv. 3313–3326)

[*He who led her into exile now holds her captive as his concubine and causes King Priam great dishonor. Thus the king, with me as his emissary, requests that you bring about his sister's return. He makes no other threats against you and declares only that he who brings her to him peacefully will find a peaceful reception. He has no desire to make claims against you, and afterwards this matter will be settled. Although he has many rightful grievances, he himself will never on this account foment any war, disputes, or large or limited aggressions against you if only his sister is freely returned to him.*]

Next, when Antenor confronts Telamon himself (vv. 3394–3405), it is Hesione's status and the marital prospects of the royal daughter that serve as the sole points of contention:

Fille a rei est, de grant parage:
Ne la deis plus en soignantage
Tenir n'aveir, quar trop est lait.
Rent li, se avras mout bien fait:

Ancor sereit bien mariee
S'il la raveit en sa contree. (vv. 3401–3405)

[*She is sister to a king, a women of noble lineage. You must not continue keeping her as a concubine, for this is an odious offense. Surrender her and you will do a fine deed. She can yet be honorably married if the king has her again in his country.*]

Perhaps it is natural that Antenor should focus exclusively on Hesione when addressing Telemon himself, but it is interesting to note that he continues to omit any mention of the other Greek crimes on his next visit at the residence of Castor and Pollux (vv. 3455–3462). As if to remove any doubt about Antenor's own role in the matter, Benoit inserts a timely reminder regarding Antenor's impeccable qualifications: "Antenor lor a dit tres bien / son message, n'i laissa rien" (vv. 3453–54).[10]

Since council scenes and direct discourse represent one of the major devices for creative amplification in the *roman d'antiquité*, Antenor's speeches are indeed significant in revealing Benoit's thematic design. Through these verbal exchanges and the motivations and judgments they expose, Benoit elaborates his own clerical gloss on the skeletal narration of events provided in the chronicles. The manner in which Benoit develops events surrounding Hesione's abduction suggests something crucial about his own reading of the historical record. It appears that the marriage theme (defined to include abduction, concubinage, and other broad marital concerns) offered Benoit a nexus for understanding the intimate links between sweeping historical events on the one hand and individual perceptions, impulses, and actions on the other. Twelfth-century learning—the study of Ovid in particular—furnished a ready set of terms and conventions for elucidating the subjective "space" of history. Only a subtle, literary sleight of hand was needed to weld together gloss and chronicle through the illusory looking glass of a (historically anachronistic) medieval *intellectio* that detextualized Ovidian passion so as to make it the vehicle of a universal knowledge to be appropriated and adapted to the historical project at hand. Undoubtedly, the marriage theme provided the most ready and comprehensive motif for explaining the intimate relation between private and political history and for uniting the twin strands of *eros* and providence into one universal vision.

Nevertheless, we still have to ask how Benoit develops his view of marital mores so that the marriage motif is more than a mere trigger for epic battles. What informs Telamon's attitude toward marriage, and how does the confrontation over Hesione begin to define Benoit's vision of the social *ethos* that drives the tragic destiny of the Greek world? How does the Hesione episode foreshadow Benoit's treatment of the more celebrated marital drama involving Paris and Helen, and how is the famous tale of abduction reconciled with his critique of flawed marriage customs? Finally, what meaning emerges from the larger distinctions between Greek and Trojan adumbrated in the contrasting marriage dramas?

Perhaps the best way to describe Benoit's portrayal of the Greeks is to compare them with the definition of Eneas's arch rival Turnus. The Greeks, like Turnus, act in accordance with a primitive "feudal" mentality. Greek society is fragmented into small "fiefdoms"—a fact made evident by Antenor's island-hopping. The hostile responses of the Greek noblemen suggest that there is a fundamental cultural gap that prevents reconciliation. Priam objects to the fact that Hesione has not been respectfully married, and Antenor makes it clear that it is Hesione's status as *concubine* that cannot be tolerated. But the intransigent and hostile posture of Telamon and his compatriots reveals that the exogamous ethic implicit to Priam's demands lay outside the customary mores of the Greeks:

> Dahé ait hui la soë [Priam's] amor
> Ne qui s'i fiera nul jor! . . .
> Quar il n'est genz cui jo tant hee
> Com ceus de la vostre contree
> Dahé ait qui les amera
> Ne qui ja o eus pais avra. (vv. 3535–50)

> [*Damned be his [Priam's] love and anyone who would ever put his trust in it! . . . For there are no people that I hate more than those from your country. Damned be anyone who love them or seek to make peace with them.*]

In this (Nestor's) reply, *amor* represents the essential basis of political reconciliation. Its meaning here, however, is restricted semantically to its older, feudal and military connotations. Associated with the notion of loyalty, *amor* in this context expresses mutual ties of affection and honor between vassal and lord or between military companions (Matoré, 179). Yet, these feudal attitudes informing Nestor's language preclude any possibility of an accord in the present circumstances. Furthermore, Telamon's own attitude toward marriage and with regard to his "possession" of Hesione conforms to the same logic: he "keeps" her, but is not moved by amorous sentimentality either as a reaction to her beauty or as a response to the nobility it represents. Telamon and the Greeks are in this way representative of tendencies that still animate twelfth-century aristocrats, while the historical dimension serves to paint the "backward" nature of such tendencies. Indeed, the contemporary (twelfth-century) fascination with Ovid and with troubadour lyricism provide an alternate frame of reference. The Greeks have not cultivated any notion of amorous affection that would provide a basis for marriage predicated not on patriarchal authority, power, and possession but on the providential operations of *eros*. Proud and aggressive, Telamon is impervious to *Amor*. In the context of twelfth-century ideas about beauty and marriage, Telamon's disdainful attitude reflects a lack of spiritual refinement and nobility.[11] From a political point of view his recalcitrant posture is not only antithetical to the diplomatic efforts of the Trojans, it is also, symbolically, a transgression against a divine law—against the binding power of *eros* whose powers of attraction have the potential of motivating ethnic alliances. As Antenor's frustrated diplomatic efforts demonstrate, Telamon, Nestor and their compatriots all display an identical antipathy toward exogamous marriage practices and toward political reconciliation (vv. 3327 ff.; 3407 ff.; 3463 ff. in contrast to Antenor's conciliatory posture, e. g., vv. 3460–01), and Benoit clearly intends us to see that the Greek mentality underlies the tragic cycles of war and vengeance that plague their universe.[12] Ultimately the dualistic vision of the more primitive "epic" ethos is shown to be incompatible with the imperatives of *eros*. Benoit alludes to twelfth-century cultural parallels by evoking, in a speech by King Priam, a conventional epic trope central to the *chanson de geste* and made famous by the roughly contemporary Oxford version of the *Song of Roland*: "Grezeis ont tort, nos avons dreit" (v. 4186). In direct opposition to the empire-forging power of love and marriage exemplified in the marriage of Eneas and Lavine, Benoit portrays the attitudes of the feudal epic (predetermined by the very rhetoric of the genre) as perpetuating absolute ethnic exclusion and the tragic determinism of the Trojan-Greek cosmos. In attributing these words to King Priam, Benoit makes it clear that the Trojans themselves are not immune to the mores that drive the Greeks, and he avoids any grossly propagandistic or nostalgic portrayal of the Trojan nation.[13] Thus, to appreciate how Benoit sought to develop the contrast of Greek and Trojan as historical models while ushering in a new heroic model appropriate to the nascent genre at hand and to its innovative ethos, we should turn to the next stage in the escalating Trojan war—a stage marked once again by a marital crime.

PARIS AND HELEN

In many respects the Paris-Helen episode (the third major marital drama prior to the outbreak of the Trojan war) parallels that of the Telamon-Hesione story, only now the abductor is a Trojan rather than a Greek. Also like the earlier episode, the Paris-Helen episode is deemed to be an immediate "cause" of the ensuing Trojan war (its fateful consequences being set in relief by the prophecies of Cassandra). In some respects, however, it differs from the Telamon episode. Most notably, it adds to the crime of abduction a second category of marital offense, that of adultery. Also, it is the refusal to properly *marry* Hesione which angers the Trojans in the Telamon episode. In the Paris-Helen story on the other hand, Paris (unlike Telamon) is not only susceptible to female charms, but he swiftly and publicly marries Helen. Benoit also builds in a larger contrast between the mores of the Greeks and Trojans. For the Greeks, marital crimes involve not only the private fortunes of individual partners but patriarchal and political prerogatives. The offense done to Hesione is justified because she is taken as a prize in war and has become the "property" of Telamon. The Trojans, on the other hand, seem less preoccupied by the act of abduction *per se* and resent Telamon's failure to grant Hesione a respectable conjugal and social status in her new homeland. By contrast, the Greeks take no consolation in the fact that Helen is promptly accorded royal status by the Trojans; her abduction alone concerns the Greeks and is judged an offense against family honor and a provocation requiring a military response. Ultimately, therefore, Benoit seems to cast a favorable light on the Trojans. Paris and Helen are wed in triumphant pomp and circumstance, and their marriage seems to stand as a corrective to the mistreatment of Hesione by the Greeks. Furthermore, the love that binds Paris and Helen attenuates the ostensible "crime" by subverting the legitimacy of Menelaus' claims (i.e., "property" claims) upon Helen. Emerging from the contrasting episodes, therefore, are competing ideologies—one in which marriage is governed by patriarchal authority in the way other property rights are governed, the other in which marriage is based on the mutual affection, honor and consent of the partners.

As Benoit recasts the famous legend in a contemporary and partisan light, however, we can also see that he maintains a certain degree of historical integrity. To understand Benoit's conscious design, we should examine closely how he develops his version of the "Judgment of Paris" and the abduction episodes. As we have pointed out, Paris not only abducts Helen, he also marries her. Dares's chronicle itself tells us as much ("Priam gave Helen to Alexander [Paris] to marry"). Benoit adds to the thematic significance of this detail by developing the marriage motif in significant ways. Benoit makes Helen a more than passive victim: she is an amorous and consenting spouse, and her active role must no doubt complicate our judgment of Paris's "abduction."

Dares's own representations may have inspired Benoit's creative extrapolations. The chronicle briefly refers to the mutual attraction of the two lovers (each one is mutually struck by the other's beauty, §10), and Helen is said to leave with Paris and his troops "not unwillingly" (§ 10). From these two details, Benoit weaves a tale of marriage compatible with amorous preludes (presided over by "Amor") and the notion of (spousal) "consent" (during the abduction Benoit says of Helen, "Ne se fist mie trop laidir, / Bien fist semblant del consentir"; 4505–6). Benoit stresses that the marriage is approved and celebrated by the larger Trojan populace.[14] He highlights the mutual consent and affection of the lovers. Paris does not treat Helen as a passive possession; he testifies to his love and fidelity in a speech in which love and marriage are tightly bound:

> "Dame," fait il, "vostre voleir
> "Sera si fait e acompli
> "Com de vostre boche iert gehi."
> Par mi la main destre l'a prise:
> Sor un feutre de porpre bise
> Sont andui alé conseillier,

Si li comença a preier:
"Dame," fait il, "ço sacheiz bien,
"Onques mais n'amai nule rien,
"Onc mais ne soi que fu amier,
"Onc mais ne m'i vous atorner.
"Or ai mon cuer se en vos mis,
"E si m'a vostre amor espris,
Que del tot sui enclins a vos.
"Leiaus amis, leiaus espos
"Vos serai mais tote ma vie:
"D'iço seiez seüre e fie.
Tote rien vos obeïra
E tote rien vos servira. . . .
Tot ço voudrai que vos voudreiz
E ço que vos comandereiz." (vv. 4730–4754)

[*"My lady," he said, "Your desires will be obeyed and granted the moment they part your lips." He took her by the right hand: together on a saddle covering of fine dark cloth the two conducted their discussions, he entreating her thus—"My lady, have no doubts, never will I love anything in this world, never again will I be a lover, never more will I think of any other but you. I have my heart so placed in you, and your love has so captivated me, that I have feelings only for you." —"Loyal friend and spouse, from this day on I too will be yours all my life, of this you can be sure. And I will obey you in all matters and serve you in every way . . . . Your will is my will, as well as your every command."*]

In responding, Helen pretends that she has no choice but to consent, but Benoit clearly wants us to see in her words the same posture of discretion she employed at the time of her abduction. She veils her own adulterous desire with consummate double-speak:

Se jo desdi e jo refus
Vostre plaisir, poi me vaudra.
Por ço sai bien qu'il m'estovra,
Vueille o ne vueille, a consentir
Vostre buen e vostre plaisir. (etc.) (vv. 4758 ff.)[15]

[*If I oppose and refuse your wishes; little good will it do me. For this much I know well, that whether I may want to or not want to, I have no choice but to agree to your kind offer and your wishes.*]

Taken in conjunction with the later account of Pyrrhus' abduction of Hermione, the Paris-Helen episode clearly defines important features of Benoit's attitude toward love and marriage. In the story of Pyrrhus and Hermione, Benoit is quite explicit about the presence of love and consent: "De nuit, *par consence*, a larron, / Prist danz Pirrus Hermiona, / Femme Orestes, si l'en mena / Ensemble o sei en sa contree; / *Prise l'ot puis e esposee.* / *Mout par l'ama de grant amor* / E mout li porta grant honor" (vv. 29596–602). The irony here is that "consence" can mean criminal "complicity" (the abduction was carried out with the help of others), or that Hermione herself was a consenting partner in an illicit love affair. The second meaning seems to prevail given the contextual references to love and marriage, but the juxtaposition of *consence* with *a larron* sharpens the irony by reminding us of the criminal implications of the lover's secret passion and pointing to the ethical ambiguities inherent to Benoit's critical treatment of marriage themes.[16]

Ultimately, Benoit goes on to resolve this engaging ambiguity. First, he reshapes the Hermione story in a way that contrasts creatively with the Ovidian. In *Heroides VIII* (Hermione to Orestes), Hermione has clearly been abducted against her will. Hermione explicitly raises the issue of willing consent ("quod potui, renui, ne non invita tenerer." 5) and suggests that conscious volition (as consent or resistance) provides the essential if limited measure of a woman's ethical conduct or "action" in such situations ("cetera femineae non valuere manus." 6). By alternatively making Hermione a *consenting* partner in an amorously motivated abduction, Benoit reorients the consent theme so that it conflicts with traditionally legitimate marital ties. The notion that love is a powerful marital principle in itself does of course appear in Ovid too and is a theme raised by Hermione's letter. The moral irony is less severe in Ovid than in Benoit, however. This is to say that Ovid's Hermione is the victim of a *malentendu*: she is caught between two valid marriages. In Ovid she is granted to Orestes not by Menelaus but by her grandfather, while Menelaus (not aware of this prior betrothal) granted her to Pyrrhus. In this dilemma, Ovid's Hermione argues that love should define the proper choice. In Benoit this Ovidian prerogative (which is to say Love's prerogative) imposes itself with even greater drama. In Benoit's text, the marital circumstances are unambiguous, and the amorous imperative runs directly counter to the prerogatives of a prior, undisputed marriage agreement. Hence, it is not the abduction, but (by inference) the marital union void of mutual love that is tantamount to imprisonment and to a transgression against a higher law. We can go on to say that the thematic link between the Helen and Hermione episodes is suggested to Benoit by an affiliation alluded to in Ovid. Hermione, while insisting that her right to return to Orestes be respected, argues that Pyrrhus is now to her what Paris was to her own mother (VII, 41–2). Hence the thematic affiliation is reinforced metaphorically by the genealogical one, and this was sure to resonate with Benoit's own pseudo-historical pretensions. In Benoit's version, however, the links between the two episodes serve to highlight the author's bold reordering of historical values. Returning to the central event, to the abduction of Helen, Benoit's courtly twelfth-century audience would undoubtedly recognize the poet's design in dwelling on amorous sentiments: love sanctioned the union of Paris and Helen while simultaneously invalidating Menelaus' claims over Helen's person. By also invoking the substance of contemporary marriage doctrines that affirmed the importance of spousal consent and affection, Benoit could both expand on the Paris-Helen episode (as reported by Dares) and concurrently exonerate the two lovers.[17] In essence, Benoit exploits the principles of (bilateral) marital consent (public and spousal) as the basis for a revision of history that legitimizes Paris's notorious crime. In so doing, he provides a foundation not only for the redemption of Paris's descendants (namely Æneas who, by his own bride conquest, initiates the critical transfer of dominion leading to Rome's ascendancy) but also, by extension, for the redemption of the Trojan myth seized upon by Benoit's Anglo-Norman patrons.

By describing the Paris-Helen relation in terms of contemporary and progressive ideas about love and marriage, Benoit complicates our moral judgment of Helen's abduction. Although Paris wrongfully usurps Menelaus's rights of possession, the love affair and ensuing marriage impose an entirely new ethical perspective that seems to transcend the Greek vision. Benoit's desire to contrast new and old in this context comes out clearly when Agamemnon attempts to console his brother Menelaus. Menelaus is cautioned against any unmanly sentiments (including love, one would suppose), and Agamemnon calls for a response worthy of the "good ol' days" of their ancient ancestors:

> Guardez que ja hom qui seit nez
> Se puisse aperceveir ne dire
> Que vos en aiez duel ne ire.
> Li preisié home del vieil tens,
> Qui tant orent valor e sens,
> Ne conquistrent pas les honors
> En duel, en lermes ne en plors . . . (vv. 4950ff.)

> [*Beware lest any man might ever see or say that you had expressed any signs of sadness or anger in this matter. The man of honor, in days gone by when men had so much valor and good sense, did not obtain their superiority by sighing, crying, and lamenting . . . .*]

As Agamemnon continues his harangue, he articulates the ethos that lies behind the cycles of vengeance that plague the Greek world:

> Mais cil a cui hom tout honor,
> Qui les granz cous a a sofrir
> E les maisniees a tenir,
> Qu'il seit povres e sofraitos,
> Or seit riches, or bosoignos,
> A la feiee guaaignanz
> Et a la feiee perdanz;
> Que sis pris creisse e mont e puit,
> Ne de bien faire ne s'enuit:
> Ensi conquistrent lor honor
> Ça en ariers nostre ancessor;
> Ensi puet l'om en pris venir. (vv. 4964–75)

[*But he who is held in high esteem and who has had to suffer harsh blows and maintain household and retinue—whether he be poor and destitute, rich or needy, whether in victory or in defeat, regardless of all these—his honor should grow and increase, while he tire not of doing good. This is how they obtained their superiority our ancestors who came before us. And this is how a man earns prestige.*]

While honor is of course a compelling motivation in the Trojan councils as well, the Greeks are clearly portrayed as pursuing the logic of honor to the point of reckless self-destruction. Benoit suggests, however, that marriage—even to an abducted spouse—offers an alternative resolution. We saw that it was above all Hesione's status as concubine that brought dishonor upon her and the royal family of Priam. Benoit makes a parallel point in the case of Helen: by dwelling on the dynamics of love and consent, Benoit implies that a happy union has replaced an unhappy union, that the abduction and new marriage represent, potentially at least, a more balanced *status quo*.

In order to articulate such a sophisticated ethical judgment, however, Benoit not only invokes contemporary marriage doctrines, he exploits the providential perspectives of a vast historical epic (creatively glossed) to provide a transcendent point of reference as the basis for a divine typology that reorients what are otherwise virtually incontrovertible moral transgressions. Like the *Enéas* poet, Benoit uses an allegory of *Eros* to provide his historical vision with a transcendent agency of causation. The island where Paris and Helen meet is the site of a cult "ancien e precios" presided over by "Venus la deesse d'amor" (4261–64). In another passing intervention, Benoit emphasizes the providential nature of the lovers' union:

En lor aé, en lor enfance,
En lor forme e en lor semblance
Les a griefment saisiz Amors:
Sovent lor fait muër colors
Tant erent bel, ne me merveil
S'il les voleit joster pareil:
Nes poüst pas aillors trover. (4359–65)

[*In their life, in their youth, in their body and their appearance, Love has completely taken hold of them. Often Love makes them change hue, so beautiful were they. Nor does it surprise me if Love's will was to assemble them both. Impossible that they be found anywhere else.*]

The emphasis on love and fate gives a mythic and providential patina to the otherwise ostensible transgressions of customary laws and mores. By tying the marriage theme to the issue of spiritual doctrine, Benoit underscores these transcendent signs of legitimacy while linking the institution of marriage to a larger providential design involving cultural dominion. At the same time, the clash of old and new perspectives and mores (as in the struggle between Eneas and Turnus in the *Enéas*) drives the underlying tragic pathos central to the Troy legend.

In his handling of the "Judgment of Paris," Benoit makes even more significant additions to his source in order to bring out the mythical aspects of the Paris-Helen marriage. In her study of the evolution of the Judgment of Paris motif, Ehrhart points out that Benoit has "considerably amplified" the three-sentence summary of the event in Dares's chronicle (Ehrhart 40).[18] Some elements in Benoit's adaptation, such as the golden apple and the individual bribes presented by the goddesses, come from classical traditions and are taken from the *Eneas* (Ehrhart 41). Benoit makes important additions to the *Eneas*-poet's version and to Dares's version, however, in his handling of the motif. In the chronicles and in the *Enéas*, Paris is identified as a hunter, not a shepherd.[19] However, Benoit makes the judgment the subject of a dream vision laced with Celtic mysticism. As Ehrhart reminds us, the dream device itself is inspired by the rationalizing approach of late-classical sources. For many new details, however, "Benoit turned to Celtic tradition, finding there a useful analogue to the Greek supernaturalism of the story's mythological origins" (Ehrhart 41–3; cf. Bromwich). In other words, Paris is portrayed in Benoit's text as a cross-over hero who undergoes an encounter with the goddesses of the Otherworld. Just as the *Enéas* poet exploits Virgil's stag hunt episode in order to integrate the mythical and ritual meanings of the Celtic sovereignty theme into his story of Eneas's quest for a bride and dominion, we are told that Paris falls asleep by a fountain while straying in the forest during a stag hunt that takes place on the first day of May. Benoit's allusion to this Celtic subtext in his account of Paris's judgment gives new ramifications to the larger saga of abduction and marriage. Helen's almost "unworldly" beauty, for example, as well as her elective desire to abscond with Paris reflect the traits of an Otherworld sovereignty goddess. These Celtic echoes cast a subtle but important shadow of historical determinism over the lovers' experiences. The sumptuous marriage between Paris and Helen now can be seen as a convenient emblem foreshadowing the providential transfer of sovereignty that links Paris to Eneas and Troy to Anglo-Norman England. The Western *translatio imperii* is emblematically fulfilled in the image of Eneas' marriage and renewed in the fortuitous union of the former French queen Eleanor and her new Anglo-Norman spouse Henri II. Paris's apparently illegitimate marriage and the fateful consequences of his actions are likewise reintegrated into the larger myth linking Troy and its descendants to the manifest destiny of subsequent European empires. Paris and Helen are portrayed as star-crossed lovers whose union is presided over by a higher providence—a providence misunderstood and fiercely resisted by the obstinately ethnocentric and patriarchal Greeks.

Perhaps it is in relation to the complex and emblematic figure of the Bride-Queen-in-exile that we can best summarize the fundamental tensions, fears, and hopes evoked in the marital dramas of the *Roman de Troie*. The sacramental grace and providential idealism that ratifies the imperial marriage and the expanded imperial dominion over Norman-Angevin territory is tainted if not undermined by the the shadow of inconstancy attached to the memory of Eleanor's first marriage to the King of France. In this respect, twelfth-century events, like the Trojan myth, project an unsettling ambiguity. Benoit responds to these typological and political ambiguities through his creative glosses of the problematic heroines and wives that animate his Greek sources. In his address to Briseida, he distances himself from popular opinion and casts Briseida in a tragic light that attenuates the shallow and external appearances of licentious infidelity. Andromache, who shares in Briseida's fate as a woman in exile (by virtue of her forceful abduction after the fall of Troy), redeems Benoit's Briseida by consummating a second and exogamous marriage that ushers in a new alliance and new peace. This dynamic and progressive revision of history and of the typology of its exemplary protagonists casts new light on the tensions underlying imperial marriages in conjunction with radical transfers of dominion.

We saw that the author of the *Eneas,* inspired quite possibly by Virgil's own reproach of Dido's neglect of fame and decorum, dramatized a providential but also highly fictional vision of *ius coniugalis* to regulate and stabilize the destructive forces of passion, revenge and war. We also saw that the author's allusion to the Trojan war as a war of private and marital revenge presided over by Menelaus allowed him to place the Trojan and Latin wars on a parallel plane: both wars are fused to a broader historical vision by the marriage theme. Thus the *Eneas* and the *Roman de Troie* are not only linked by the Trojan pedigrees of Eneas and Paris, they also seem to join in exploiting the marriage theme as a focal point for unifying both their own vast narratives and their own historical perspectives.

TWELFTH-CENTURY MARRIAGE REFORMS

Our analysis of the *Roman de Troie* shows that Benoit's narrative, far more than an exercise in translation and anachronism, is a purposeful and expansive reflexion on marriage practice. We can only wonder how Benoit's imagination allowed him to recapture from between the lines of such stark Latin chronicles the dark and tragic pathos that characterizes the Greek psyche. For our purposes it is particularly interesting to have observed that Benoit, in seeking points of resonance for his own audience, portrayed ancient Greek society as one ravaged by civil strife and "glossed" this tragic vision in terms of aristocratic marriage customs based on a model of economic conquest and exchange. Originating in the sphere of domestic violence and extending into the theater of epic history, before circling back again to the world of private affairs, the course of history in Benoit's narrative is consistently driven by marital failures and successes. The Greek and Trojan world, according to the logic of Christian typology, dimly but forcefully adumbrates a universal and divinely sanctioned vision of legitimate marriage practice that transcends the prerogatives of common law. As *exemplum*, the Greek and Trojan world is ambiguous. The pagan world remains deprived of God's revelation, but the course of history itself, the implosive destruction of the Greek nation in contrast to the ultimate ascendancy of Rome as the New Troy, is revelatory to Christians in its own right.

Obviously, contemporary views and controversies relating to marriage captured Benoit's attention, and the tragic and epic dimensions of his Troy narrative underscore the depth and scope of philosophical and social considerations that were now seen to radiate from this central motif. The *Roman de Troie* deserves special recognition from students of medieval literature in this regard. It demonstrates that in the middle of the twelfth century marriage had become a crucial literary theme and intellectual preoccupation. It also reveals that the interest in adulterous and erotic motifs is part of a larger interest in the divine or "sacramental" dimension of secular marriage—a dimension that links epic marriage conflicts to the fundamental outlines of European history. It is therefore surprising that critical treatments of love and marriage in Old French literature have long oscillated primarily between references to troubadour lyric and the notion of *fin'amor* on the one hand and Augustinian morality on the other. If the *Roman d'Enéas* and the *Roman de Troie* accurately anticipate the development of the Old French *roman*, then classical and Celtic literary traditions seem to have attracted at least equal attention. The mythical and marvelous motifs in these traditions as well as the social and marital mores which subtend classical epic would certainly have invited reflexion, for in the twelfth century Christian marriage doctrine entered into an unprecedented stage of innovative development which would allow for a remarkable syncretism.

Marriage controversies loomed large in twelfth-century culture and thought, and the nature of medieval marriage fictions indicates not only that twelfth-century authors were cognizant of such controversies, but that they looked beyond Roman and patristic precedents. Jean Markale has pointed to the alterity of Celtic marriage mores in relation to Christian and Roman attitudes, and the popularity of the Tristan legend as well as the *Lais* of Marie de France are convincing indicators that Celtic legends, although leaving no direct trace in canonical texts, provided an alternate tradition for imagining the dynamics of love and marriage. Even in the *Canterbury Tales*, Chaucer's compendium of medieval culture, we discover that reference to Celtic lore was synonymous with a partisan perspective on sexual and marital mores. At the beginning of her tale, Chaucer's famous "Wife" nostalgically recalls the days of Arthur and laments the disappearance of jolly fairies who have everywhere been exorcised by English monks and friars as misogynist as they are ubiquitous.[20] For Markale, Celtic attitudes provided not only an important corollary but also an initial basis for the ethos of *fin'amor* which he defines as representing "a real rebellion against established ideas" in a Christian period (265). But any strict opposition between Celtic norms and those of the "Mediterranean mentality" which Markale says is "soaked in Roman jurisprudence and severity" (261) reduces the question of *fin'amor* and the history of medieval marriage doctrine to an oversimplified polemic, just as Chaucer satirically exaggerates the opposition between the Wife's licentious attitudes and the extremes of patristic misogyny as the Wife and her fifth husband rush to their final explosive confrontation. We have seen too (chapter one) that the Old French *Tristan* tales do not simply exalt illicit love in opposition to conventional marital propriety. In light of the perspectives gained by a close reading of Benoit's *Roman de Troie*, it is increasingly clear that the tragic trajectory of the *Tristan* tale testifies to the sacred and indomitable power of amorous passion while simultaneously highlighting the need (for individuals and society at large) to reconcile libidinous desire with marital practices. At the same time, we are forced to ask what precedent provided a model for an integrated vision of erotic passion and civil marriage. Benoit's work reveals that emerging reforms in the area of Western marriage law and marriage theory provided an impetus for this broader synthesis. In order to appreciate subsequent marriage fictions in the Old French *roman*, we need first, therefore, to have a fuller understanding of the content of these reforms and of the principal motivations and circumstances surrounding their development.

In the eleventh and twelfth centuries, medieval attitudes toward marriage included dogmatic practices and opinions informed in some cases by religious ideology, in others by secular customs, and emerging marriage doctrines negotiated both traditions and relied on both pragmatic and speculative approaches in response to what historians agree was a noteworthy paucity of established precedents and principles for defining a Christian doctrine of marriage.[21] What precedents proved influential and how were they appropriated or defined by medieval thinkers? Why did marriage speculation figure prominently as a theme in secular fiction rather than remaining confined to Latin scholarship—to canon law studies, and patristic theology? What legal and theological developments in the area of marriage doctrine allowed twelfth-century writers to envision amorous and marital themes in such sweeping dimensions involving psychological, legal, ethical, national and mystical elements? What revered if scant precedents (precedents from biblical scripture supplemented by patristic writings and studies of Roman law) would have provided a basis for such an elaborate and imaginative vision of the role of marriage in relation to the secular orders and to traditional social custom?

One point made clear by modern studies of medieval marriage is that marriage has a unique status alongside other Christian sacraments:

> The central fact that most differentiates marriage as a sacrament from other sacraments in the writings of the canonists and theologians in the first half of the twelfth century is that marriage existed as a social and legal institution, and always had, regardless of what Christian thinkers might say about it. It had a life of its own apart from Christianity. (Colish 628)[22]

In the form of rite, bond and institution, marriage practices had developed apart from Christian doctrine and were defined by social customs and structures that existed independently of theological interests in grace and salvation. Even within the confines of theological doctrine, conflicting viewpoints existed as to the moral and sacramental status of marriage, and deeply ingrained attitudes toward concupiscence and sexuality were obstacles to the Church's capacity to integrate an ethical theory of marriage into its wider pastoral mission.

However, moral caution or reproach did not necessarily result in simply benign disinterest with regard to secular marriage practice. Some organized sects voiced outward contempt for marriage while many in the religious orders felt free to marry or associate with concubines. So it is that the history of medieval marriage reform has its critical starting point not in the philosophical, artistic, and theological movements of the twelfth century nor as a reaction against any perceived ethic exalting free love, but in the eleventh century. It is in the closing decades of the eleventh century—years recognized by Duby to be a watershed in the development of Western marriage law (*Love* 13)—that Ivo and other church officials in France actively began to outline a code of marriage law (based on citations from Roman law, scripture and the Church Fathers), not with the immediate purpose of imposing secular reforms but out of the need to combat religious heresy and to condemn the abuses of Nicolaism within the orthodox religious orders (Duby *Love* 15–16; *Two Models* 18–20).

Ivo's repudiations of heretical views on marriage reflect the temper and tenor of earlier doctrinal controversies over marriage in the patristic period. Fundamentally, these earlier debates and schisms were generated by moral concerns based on varying degrees of asceticism and laxity as disagreements arose over the moral sanctity of conjugal life. Some Fathers would express strongly anti-matrimonial views, while others attempted to extol marriage as being morally superior to virginity (Godefroy 2077ff.). Over time, however, it is possible to discern a growing consensus around moderate views defining marriage as a blessed state (albeit one spiritually inferior to that of religious celibacy).[23]

In Ivo's time, however, it was more than a question of reiterating orthodox doctrine. There was an urgent need for institutional reforms within the Church hierarchy—the need to curb clerical promiscuity and the resentments it fostered. Yet these abuses themselves were not simply a sign of moral laxity; they were part and parcel of a continuing controversy within the Church relating to marital practices. The controversy was aggravated by the lack of any positive and univocal authority (or set of authorities) which could provide for clear theological and legal formulations as the basis for an institutional consensus (Noonan, *Consent*; Sheehan; Godefroy 2068; Le Bras 2125–6).[24] In the area of marriage practices, therefore, the elaboration of systematic theological doctrine and sacramental teachings would come only after the effort to articulate more pragmatic reforms.[25]

At the same time that the claims of heretics and calls for reform underscored the need for coherent marriage laws, new sources of legal study (to complement biblical scripture and patristic doctrine) were being made available through rediscovered tracts from Roman law (Duby *Two Models* 20; Le Bras 2123–4). All of the historical evidence suggests that up until this time, medieval marriage practices had been governed by local customs and family political and economic interests that remained unconstrained by any broader social or philosophical consensus—a doctrinal consensus that up to the middle of the twelfth-century remained as unfinished business in Western Christendom. Paradoxically, the very paucity of authority that often limited the efforts of Ivo to *ad hoc* and politically expedient formulations would ultimately leave the door open to a wide range of theoretical discussion. In the context of the intellectual developments and vibrant ecclesiastical, legal and spiritual reforms that characterize the twelfth-century, the stage was set for dramatic developments in the area of marriage doctrine—developments which crystallized during the middle of the twelfth century around two successors to Ivo of Chartres, Gratian, a prominent canonist, and Peter Lombard, a prominent theologian.

Once Ivo of Chartres had clarified the Church's position that marriage represented a blessed stated, the primary concern would no longer be the proposition "to marry or not to marry" but to define and come to agreement on the nature of the conjugal state as a Christian institution and Christian sacrament. As a result, "the definition of marriage and of marriage formation" became "the single biggest debated question raised with respect to this sacrament in first half of the twelfth century" (Colish II: 658). Many questions were raised. What is the sacramental value of conjugal life, and what makes it an effective symbol and vehicle of grace? What criteria should inform judgments regarding 1) the *legitimacy* of a given marriage contract, and 2) the *validity* of the conjugal bond itself (the nature and integrity of the spiritual bond underlying the customary rites, gestures, and exchanges)? The question of legitimacy called for legal precedents and formulations, that of validity for moral and theological ones. Yet, legal and religious traditions would ultimately be forced (creatively and imaginatively if necessary) into essential conformity by the need for a consistent orthodox logic as the foundation for a univocal ecclesiastical policy on marriage. Thus, despite distinct precedents and aims, both legal and speculative developments are influenced by a larger intellectual imperative and exchange.

As celibacy and virginity were more a focus of attention than marriage for most Church authorities and for the religious orders generally, marriage law developments came to depend upon references to Roman and Germanic customs (Brundage 187). But the influence of Papal interests was manifest in the tendency to invest a greater share of marital authority in the individual as a means of undermining the paternal, familial and economic prerogatives privileged by secular custom and serving to insure aristocratic autonomy (Sheehan 8–9). Indeed, we have shown how the authors of the *Eneas* and *Troie* provided apposite glosses in this regard not only by elaborating various marital conflicts but also by contrasting customary feudal marriage practice and precedent with a transcendent vision of love and marriage based on the providential agency of Venus and the operations of erotic attraction. Hence, as we shall see, the notion of spousal consent articulated in Roman law texts came to provide a pivotal concept for later doctrinal developments that sought to define marriage as a sacrament and vehicle of divine grace. At the same time, however, the notion of consent was a legal principle which had its own roots in secular customs and not in a logic that inherently conformed to or was explicit in Christian teaching.[26] Thus to some extent it will not be surprising to see theologians attempting to *harmonize* the edicts of God's eternal will with the key concepts highlighted in canon law—such that the marriage sacrament might be ratified by its appeal and conformity to historical precedent as an alternate source of revelation. But in searching for a spiritual element which might serve to validate the link between the sacrament and the individual soul, theologians had to navigate a complex array of amorous affects and desires. In this context, the notion of marital *affection* to be found in Roman law texts provided a precedent for envisioning a spiritual operation as the basis for a concept of conjugal union that goes beyond the purely contractual function of individual consent. At the same time, however, there was bound to be slippage between such a vaguely subjective and affective (if specifically legal and religious) notion and a larger sphere of secular sentimentality informed by individual experience and the salient Western (erotic) myths that inform enduring cultural attitudes toward love. Nonetheless, religious formulations would seek to incorporate the notion of marital affection into a model of marriage that did not promulgate or perpetuate sinful or anarchic tendencies. As we focus on particular doctrinal developments, therefore, we should bear in mind 1) the lack of complete legal and patristic authorities and the unusual opportunity this provided for inventive formulations; 2) the cultural traditions (non-legal and non-patristic) available as alternate arenas of speculation; 3) the simultaneous constraints imposed by religious and ethical imperatives.

The basis for defining the sacramental value of marriage goes back to the New Testament itself, and to the early Church Fathers. It is the apostle Paul (*Ephesians* V, 22–33) who refers to marriage as a "*sacramentum*" and attributes to it a symbolic significance by analogy with the mystical union between Christ and his Church.[27] Did this symbolic resemblance clearly attest to a positive sacramental value in the conjugal bond itself? Paul's text provided the basis for a decree to this effect and for its triumphant assertion in the twelfth century. However, the question remained contentious (Le Bras 2177; Brundage 140). The apostle's own famous dictum "it is better to marry than to burn" (1 Corinthians VII, 9) set the tone for an opposing tradition. Peter Abelard and other early twelfth-century thinkers followed Augustine and Jerome in seeing marriage as a kind of *pis-aller* which was merely a necessary remedy for sin. Abelard asserts that marriage is to be counted as a sacrament, "yet he can find nothing in the relations between spouses that signifies divine grace, his general definition of sacrament" (Colish II: 640–2).

At the same time, however, the Manichean heresy along with its extreme anti-matrimonial doctrines, encouraged unusually optimistic views of marriage among a number of twelfth-century theologians. Many thinkers in France "agreed in rejecting Augustine's view of marital sex as inherently sinful. If the spouses act for a conjugal good—with a procreative intention—then no sin attaches to them" (Colish 2: 659). In remarks that implicitly target the Cathars, Abelard stressed that as an institution marriage is a *res bona*. The sanctity of Christ's union with the Church testifies to the positive value of the institution and provides the basis for its status as a sacrament drawing spouses to holy ends—a vehicle for divine grace (Colish 650–1; Duby *Love* 17). In opposing the Cathars, Abelard drew attention away from the notions of concupiscence and sin limiting the value of marriage to a mere moral "remedy" and affirmed the spiritual benefits of conjugal life. Hugh of St. Victor had preceded Abelard on this path when, harking back to Augustinian tradition, he asserted that the sacraments do not merely signify, they also effect what they signify (Colish 1: 49; 2: 524). This view had its own precedent in the New Testament, but the tone of the text is implicitly negative: for Paul grace was necessarily a component of Christian conjugal life, because without grace the individual would not endure the vicissitudes of marriage, nor successfully fulfill the duties of marriage (Godefroy 2114; Colish 1: 51; 2: 641–2). Hugh's spiritual orientation is obviously less severe, more mystical and more optimistic. He defined the sacramental value of marriage in terms of the inner grace it communicated (Colish 2:524). He reiterates the notion that marriage is a remedy for sin, but attaches, as Colish has pointed out, a positive moral dimension by adding that the sacrament is also given for our instruction and growth in virtue (Colish 2: 643–4). This represents a decisive departure from conventional patristic attitudes (stressing the virtues of procreation) as well as an alternate field of speculation (in contrast to canonical studies concerned with purely judicial and external aspects of marriage formation). In fact, Hugh's formulations reflect the extent to which Church doctrine began to incorporate the principles of natural law as a response to excessive and heretical forms of asceticism and as an alternative to the lack of explicit scriptural authority. Hugh resolves the moral ambivalence surrounding the issues of love, concupiscence and copulation, by arguing that marriage precedes sin as a natural institution (*officium*). Only in its secondary role is marriage defined (as it had been by the early Church) as a remedy for carnal desire, as a "compact of love" which "excuses what belonged to weakness and evil in the mingling of the flesh." In its primary role, the conjugal union is a universal institution rooted in the natural and divinely sanctioned laws of procreation (a gloss of *Genesis* I, 28; see *De sacramentis* II. 11; 2). The moral ambivalence surrounding the physical aspects of consummation are resolved in turn according to orthodox Christological analogies (Duby *Love* 17). Marriage, in the intercourse of flesh, typifies the union between Christ and the Church through Christ's assumption of flesh such that "the sacrament of Christ and the Church could not be where carnal commerce had not been," although true marriage can exist in purely charitable terms without carnal commerce (*De sacramentis* II. 11; 3; p. 326). Hugh's positive formulations also involve a respectful posture toward women as worthy partners of man (*De sacramentis* II; 11; 2; cf. *Corinthians* VII), and the marriage union is exalted as a sign of the love that unites God to the rational soul internally through His grace (*De sacramentis* II; 11; 3).

Hugh's perspective would provide the basis for a growing unanimity among Church writers as to the sacramental value of marriage in Christian life despite lingering disagreements as to the exact value of marriage as a sacrament given its origin in natural as opposed to explicit New Testament law (Colish 2:629). Furthermore, while some writers defended marriage by reference to God's command to procreate, others focused on the mystical union of Christ with his Church (Le Bras 2148). Even Abelard joined other theologians in acknowledging marriage to be a Christian sacrament despite his otherwise hesitant posture toward conjugal life—a probable result of his own turbulent affair with Heloise (Colish 1: 51 and 2: 641–2). Even in spite of his devotion to the virtues of celibacy, Abelard expressed unusually naturalistic views of human sexuality. Abelard and a small number of his contemporaries defined love as a natural and positive force, as an outlet for sexual urges with its own intrinsic worth (Brundage 187 and 197).

It is somewhat paradoxical that Abelard's liberal moral posture toward sexual indulgence did not translate into a positive view of married life just as the response to heretical asceticism did not take the form of moral laxity. Indeed, we have seen that natural law perspectives were thoroughly assimilated and subordinated to elaborate Christological analogies in the formulations of Hugh of St. Victor. Both Hugh and Peter Lombard rehabilitated marriage as a positive sacrament by emphatically *idealizing* the nature of conjugal life (Duby *Love* 11 and 17). Peter stresses the essential importance of charity in marital relations and, evoking Mary and Joseph as exemplary figures, argues that physical union and procreation are not indispensable features of a valid conjugal union. It is this kind of synthetic formulation that provided a solid basis for spiritual evaluations of conjugal life as a divine and mystical union. Like Hugh, Peter successfully assimilates natural motivations into his spiritual scheme of marriage while at the same time harmonizing his views on the marriage sacrament with a broader Christological perspective. For Peter Lombard, writing in the middle the twelfth century, the sacramental value of marriage depends upon the union between Christ and Church and is effective on two distinct but analogous planes: a bond of charity joining Christ and his Church joins the two spouses in a spiritual union with charitable ends, while the corporal union in marriage is prefigured in Christ's status as a mortal individual in his union with the Church by nature (Colish 2: 651–2).

The strict logical implications of these theological formulations had important consequences for how thinkers viewed the role played by the conjugal bond in relation to the spiritual perfection and salvation of individuals and larger Christian society alike. Hugh of St.Victor argues that marriage and the other Christian sacraments were instituted for the sanctification of Christians and work in the soul for the recipient's *humilitatio*, *eruditio*, and *exercitatio*. Peter, writing implicitly against the Cathars, stresses that even after the Fall, marriage as an institution has a positive value, is a *res bona* which draws spouses to a holy end in accordance with its "sacramental virtue" (*res sacramenti*) (Colish 2: 524 and 650–1). Such theological formulations clearly abetted papal efforts to remodel European society along the lines of a united *civitas dei* (Noonan, "Consent" 430–1), but they also established new prerogatives—prerogatives that would compete with aristocratic customs for control over one of Western society's most pivotal social institutions at a time when marital conquest rivaled military confrontation as a primary arm of political and territorial expansion. These more pragmatic implications of theological doctrine can perhaps best be approached according to the ways in which formulations of the marriage sacrament converged with key concepts of canon law.

If the Church were to legislate and preside over marital affairs, its judicial principles would require the sanction of theological doctrines just as the recourse to natural law on the part of theologians would logically validate the principles underlying emerging canon law formulations. Like their counterparts in the field of systematic theology, legislators such as Gratian circumvented the lack of direct scriptural authority by recourse to an alternative but no less transcendent authority—that of natural law as distinguished from the arbitrary decrees of positive law that account for competing forms of social custom and mores:

Une législation peut en effet rester humaine parce qu'elle émane d'un instinct naturel sans être pourvue d'une sanction immédiate de la part de Dieu, comme l'union de l'homme et de la femme, l'hérédité, l'acquisition libre des biens sans maître. Ces derniers éléments sont bien humains, puisque non contenus dans la Loi et l'Evangile, comme le voulait Gratien, mais naturels, puisque provenant d'un instinct naturel à l'homme. (Le Bras et. al. 389)

[*A law can effectively be held as humane because it derives from a natural instinct without being accorded any direct sanction by God, as is the case (for example) in the union of a man of woman, as regards heredity, or the free acquisition of goods without a master. These cases are clearly humane, since they are not provided for in biblical law or the Gospel, as Gratien would have it, but in nature, since they derive from instincts innate in man.*]

The appeal to a transcendent authority resulted in new imperatives. One feature of Church doctrine which fit logically with the emerging emphasis on the mystical significance of the marriage bond was the principle of indissolubility. This principle, which was a central feature of emerging marriage doctrine, came into direct conflict with the customary practices of the lay aristocracy. Secular lords naturally defended their right to marry, repudiate spouses and remarry at will according to personal, economic, and political interests (Duby, *Two Models* 7–8). Once theologians defined marriage as a mystical union, however, the principle of indissolubility could ultimately restrict even the discretionary authority of Church officials who selectively relied on consanguinity impediments as one more way of controlling aristocratic marriages (Duby, *Two Models* 21–2). Nonetheless, the growing insistence on the sacred and indissoluble nature of the marital bond ultimately proved advantageous to the Church since the consanguinity impediment (although it could be applied to hinder the practice of endogamy) was the restriction most readily invoked by kings and princes seeking to dissolve a standing marriage (Duby, *Two Models* 45; Noonan, "Consent" 429). Whereas consanguinity impediments provided an enduring pretext for repudiating a marital partner, the principle of indissolubility offered the Church a means of limiting aristocratic prerogatives. Finally, given the logic of emerging theological doctrines, the principle of indissolubility gradually gained greater and greater precedence.[28]

The question of indissolubility did not, however, directly resolve the critical twelfth-century disputes concerning the nature of legitimate marriage formation. This controversy brought into the foreground two critical principles adopted from Roman law, namely the principles of marital consent and marital affection. These principles were to become critical cornerstones of emerging doctrine, and their underlying logic imposed new parameters for evaluating marital legitimacy. The Roman law notions of consent and affection placed new emphasis on the element of individual choice and volition in the legitimate establishment of conjugal bonds and provided a legal basis for spousal autonomy while undermining any rational defense for arbitrary repudiations and for the imposition of external (social, family, paternal) authority. Was this emerging shift in marital doctrine a result of conscious partisan strategies or an evolution propelled by much larger historical and cultural precedents? Duby's analysis reveals that the logic of legal (Roman law) precedents was exploited to allow for increasing spousal autonomy as a reaction against the use of paternal or social coercion. At the same time, however, the notion of consent potentially conflicted with the Church's own traditional regard for the principles of paternal guidance, social cohesion and sexual restraint (Duby, *Knight* 150). Despite this underlying conservatism, the logic of consent theory became even more compelling as theologians (grasping for authoritative precedents that would support a sacramental theory of marriage) exploited the complementary relations between the legal concept of individual consent and the subjective aspects of individual volition necessary to the operations of grace associated with the sacramental properties of the conjugal bond (Duby, *Love* 11). In elaborating legal, theological and natural law doctrines that privileged the role of subjective dispositions and individual consent, however, did not the Church simply open a door to new kinds of arbitrary aristocratic marriage practices and provide a solid basis for a radical form of secular autonomy in a relationship central to social cohesion and which it sought to imbue with a heightened religious sanctity?[29] We will see that marriage fictions in Old French narratives addressed to secular audiences dwell on these very ambiguities by contrasting and defining subtle shadings of licit and illicit amorous and marital conduct while exploring the legal, ethical and philosophical perspectives coalescing around contemporary marriage doctrines and controversies.

Evidence for the direct influence of contemporary marriage reforms on fictional narratives is perhaps best demonstrated by the preoccupation in both domains with the specific problem of marital legitimacy. In particular, contemporary thinkers pondered the ultimate and essential basis of marital legitimacy, and the objective act that ratified marriage. This was a sticky problem for a culture accustomed to relying on formal public rites and gestures in the establishment of critical feudal contracts—contracts similar to those confirming a marital union and which involve oaths of trust, duty and fidelity.[30] Emerging twelfth-century marriage doctrines only complicated matters by emphasizing and elaborating increasingly subjective, and therefore invisible, components of the legitimate marriage contract. Traditionally questions of legitimacy and litigation in marital affairs centered on consanguinity impediments and proofs relative thereto. Historians agree that now, however, whole new domains of inquiry and speculation had opened up leading to intense debate over the nature and specificity of the "act" underlying marital legitimacy.[31] In light of the attention paid to such controversies, I would like to examine briefly the nature of the two key concepts, consent and marital affection, that (in conjunction with the principle of indissolubility) were the essential elements of emerging marriage doctrine. Afterwards we will turn our attention back to the treatment of marriage in Old French narratives.

CONSENT DOCTRINE

When we think of the emphasis placed on marital consent in the twelfth century, it is, intuitively, against the backdrop of aristocratic customs. Georges Duby, for example, found it convenient to describe the development of twelfth-century marriage doctrine according to two opposing "models" of marriage, one "lay" the other "ecclesiastic." The prerogatives privileged by the "lay" model reflect the economic interests and patrimonial authority of aristocratic families. Marriage involves real economic as well as symbolic exchanges, and marriage arrangements are designed to further the economic, political, and territorial aspirations of aristocratic families. Monogamy and fidelity were not necessarily revered as aristocratic individuals sought to determine freely their marital fortunes as dictated by the contingencies of honor, childbearing, and political expedience (*Medieval Marriage* 4–7). It made tolerable allowances for the practice of adultery (by men), but abduction was abhorred because it posed a serious threat to property rights and to aristocratic interests in public order and control generally (*Medieval Marriage* 4 and 7–8). The "ecclesiastical" model, on the other hand, emphasizes the sacramental and consensual aspects of marriage as a spiritual and sacred union between two equal individuals. It condemns adultery for both sexes and condemns repudiation (unless warranted by grievous moral transgressions) (*Medieval Marriage* 16–17). Duby's two models shed a revealing light on salient aspects of the portrayal of aristocratic marriage conflicts in both the *Enéas* and the *Roman de Troie*. Eneas, for example, is condemned by Turnus and the queen as being an abductor. In the *Roman de Troie*, the abduction of the Trojan Hesione and that of the Greek Helen lead to acts of serious retaliation. At the same time, however, the abductions are condoned (in the cases of the Trojan heroes Eneas and Paris) by virtue of the *consent* expressed explicitly or implicitly by the abducted bride. Likewise, the abduction of Hesione is condemned by Benoit precisely because Telamon compels her to remain as his concubine. Hence, marital consent serves to insure proper and legitimate bonds and acts as a check against what are depicted as arbitrary if customary feudal practices.

Yet this ecclesiastical concept of consent conceals a significant and problematic ambiguity because it is expressed essentially as volition but proceeds from a transcendent order and agency. The attraction between Dido and Eneas for example is an expression of mutual desire but one which is depicted as being clearly at odds with the providential order. This ambiguity has its roots in the concept of consent itself which in no way originates as an ecclesiastical concept. On the contrary, the notion of consent (although characteristic of Duby's ecclesiastical model) descends directly from Roman practices. In the period of the early Church, the populace was already steeped in a marriage practice predicated on the notion of consent (Gaudemet 57). In the twelfth century, the notion is appropriated by ecclesiastical authorities (jurists and theologians) from Roman law texts. In both Roman and ecclesiastical doctrine, the notion of consent is likewise hardly distinguishable from the aristocratic ideal of free will. This fact harmonizes with Dumézil's observation that such an ideal (although in competition with corporate forms of social and paternal authority) has verifiable origins in specific strata of Indo-European hierarchies and that these ideals explicitly account for certain features of Western marriage custom. Tertullian, for example, asserted that it is *voluntas* which perfects the marital union, and Gratian adopts a similar formulation from another source (i.e., "Matrimonium enim non facit coitus sed voluntas, et ideo non solvit separatio corporis sed separatio voluntatis").[32] Several decades later, however, we find Chrétien de Troyes attributing the following sententious formulation to Guinevere in a well-known passage from *Cligès* in which the legendary queen advises the naïve lovers Alexander and Soredamors:

> Or vos lo que ja ne querez
> Force ne volanté d'amor.
> Par mariage et par enor
> Vos antraconpaigniez ansanble (vv. 2286–9; emphasis added)

> [*Hence I advise that you never seek willful or coercive love. Better you should live together in marriage and in honor.*]

We also find that twelfth-century theologians who assert the primacy of consent in the formation of marriage avoid the term *voluntas*, employing instead *consensus* or *foederatio*.[33]

Thus, the notion of consent has its direct precedents in Roman Law, but it undergoes important transformations. As Gaudemet points out, the notion of indissolubility central to sacramental views of marriage reoriented Christian references to the principle of consent by making the marriage bond into a sacred and *permanent* obligation. Obviously, this obligation qualified any notion of truly individual autonomy. At the same time, however, theologians never explicitly distinguish between old and new notions of consent:

> Mais la doctrine chrétienne tient le mariage pour indissoluble. Elle ne peut donc plus, comme le faisaient les juristes romains de l'époque païenne, affirmer que le mariage cesse dès que l'accord des volontés disparaît. Le consentement n'est plus envisagé comme le soutien permanent du mariage. Il s'agit d'un consentement qui établit un statut. Une fois instauré, celui-ci, bien que conventionnel par ses origines, échappe à la volonté des conjoints qui ne peuvent le faire cesser. Consentement continu en droit

romain classique, consentement initial dans la doctrine chrétienne, dit-on parfois. La distinction est séduisante. Elle rend compte de la différence entre les deux régimes. Mais aucun texte ne prouve qu'elle ait été proposée ni par les juristes ni par les auteurs chrétiens. (Gaudemet 59)

[*But Christian doctrine considers marriage an indissoluble bond. The doctrine can no longer be construed (as it was by the Roman jurists of the pagan era) to state that a marriage dissolves once free consent has disappeared. Consent is no longer seen as the permanent essence of the marriage bond. It is a question of a consent which serves to establish a formal status. Once established, although based on an initial agreement between partners, the bond no longer depends on the will of the spouses, who can no longer dissolve it. Continuing consent in classical Roman law, initial consent in Christian doctrine—this is the way some have defined it. The distinction is enticing. It conveniently describes the difference between the competing doctrines. But no text provides any evidence that such a distinction was put forth by either the jurists or the Christian authors.*]

To expect to find any such explicit distinction is, I think, to misunderstand the assumptions that inform twelfth-century concepts of marital consent. First of all, both canonists and theologians invoked Roman law as an authority for their own doctrines, hence an illusion of continuity was more important than any sophisticated form of verbal contrast. Secondly (from the vantage point of twelfth-century writers), past precedents involved overlapping Roman and patristic texts. Hence, an inherently divine principle was revealed in Roman law concepts, and their interpretations should therefore be germane to contemporary orientations. Finally, one can suppose (given that the conjugal bond is supposed to be instituted by God for all time) that the underlying assumptions of twelfth-century Christian authors and their contemporary "glosses" governed the comprehension and assimilation of the principles in the first place. Since a legitimate marriage bond was regarded as involving an infusion of divine grace and a transcendent sanction, there would be no reason to distinguish between initial and continuous consent. Twelfth-century notions of consent ultimately came to rely on the logic of spiritual engagement and not legal objectivity or formal pledges as such. The notion of consent was (for theologians and most canonists) pregnant with evident meaning and logic, for it was a convenient and agreed upon term for denoting the spiritual (subjective and charitable) engagement necessary to the sacramental operations of marriage. This logic of subjective engagement was reinforced by the growing interest in the psychology of volition and intention as fundamental features of Christian faith. This is precisely how Colish defines Peter Lombard's position with respect to Gratian. While Peter follows Gratian closely, he emphasizes the pastoral and moral aspects of marriage as a sacrament and is concerned with the way in which the sacraments are internalized in the spiritual lives of the people who receive them (Colish 1: 82). Hence the logic of Lombard's *consensual* viewpoint in no way defines itself against the backdrop of Roman tradition as such. Rather, it assumes a broader view of ethical theory in which virtue is defined as a function of subjective volition and "intentionality" (Colish 1: 84):

> Peter Lombard subscribes to the *consensus* view that sees *intentionality* as the essence of the moral act and as a description of the moral status of the moral agent even in the absence of the act. (Colish 2: 480)

The logic that informs the theological view of consent goes back not explicitly to Roman practices but to the ethical theories of early patristic writers (e.g. Augustine and Jerome; Colish 2: 473).

Of course this emphasis on spirituality would make for superior ethics but not necessarily for pragmatic adjudication. Perhaps it is this deeper clash of motives and logic that accounts for the intense scrutiny given to the problem of defining the specific moment and gesture ratifying the formation of legitimate marriage ties and for the emergence of a crucial debate (c. 1150–1170) between an "Italian school" (to which Gratian adheres) that isolates the *copula* as the objective act consummating or "perfecting" the marital bond and a "French school" that defines wilfull consent (however demonstrated) as the essential legitimizing requirement that validates the marital bond and its sacramental efficacy (Le Bras 2149–57).[34]

The "Italian school" or consummationist party was concerned with defining a uniform code of marital law. Its partisans argued that if canon lawyers and ecclesiastical officials were going to be served in actual cases of adjudication, any such code would have to provide a practical means for evaluating and *verifying* (objectively) the consummation of marital bonds and engagements. At the same time, leading theologians in France ("French school") were preoccupied with affective concepts that would define the legitimacy of conjugal bonds in accordance with the operations of a religious sacrament involving the spiritual state and willful disposition of the believer. Ultimately, the theological orientation would conflict with that of the strict canon law doctor, the latter seeking legislative precepts while the former sought to imbue marriage practices with a deeper transcendent mystery (*sacramentum*) based upon principles that even some canon law specialists (such as Gratian) recognized to have a transcendent authority because based upon the prerogatives of "natural law" (i.e., emanating from instinctive bonds and imperatives that take precedence over "positive law" statutes and arbitrary social customs).[35]

The position advocated by the "French school," which privileged the "consensual" as opposed to "consummationist" view of marriage formation, was more widely accepted and prevailed in the end (G. Le Bras 2156; Sheehan 14–15). It is perhaps hard to say with certainty why this view prevailed, but it is not surprising that it did. For one thing, the position of the "Italian School" could be seen as flawed to the extent that it failed to satisfy the need for practical proofs which preoccupied legal scholars in the first place. There was an inherent contradiction in the fact that the "consummationists" sought to base marital legitimacy on an objective and publicly verifiable demonstration, rite, or gesture on the one hand, while simultaneously insisting that the sexual act (the most clandestine of acts!) is what truly validates a marital bond.[36] Of course, the "consummationist" view was also flawed because it conflicted with the patristic iconography that associated Mary and Joseph with an idealized model of marriage involving a purely spiritual bond, and because it based the validity of the marital state upon a purely physical act while entirely neglecting the question of the spiritual state of the partners (Colish 2: 630). Secondly, the logical orientation driving the consensual view fit readily, as we suggested, with the mystical aspects of a sacramental theology emphasizing the spiritual state of married partners as well as the operations of divine grace effected through the sacrament.

Although the consensus view prevailed in the end, it was plagued by the subjective nature of its logic. The objections raised by the consummationists highlighted the inherent ambiguities. For legal scholars the physical union could at least be seen to provide a single objective act marking the true consummation of the marital bond at a specific moment in time and free from the obscurities surrounding subjective intent, amorous affection and spiritual engagement. In fact, their objections to the consent school anticipate the kind of ethical and subjective complexities which make twelfth-century marriage doctrine a focal point of controversy and speculation. Those who objected to the consent position would argue, for example, that a domestic partnership (i.e., what we call "living together" by consent) would effectively be equal to a legitimate marriage. They also anticipate the way in which the consent doctrine will complicate attitudes toward clandestine marriage vows by arguing that the consent view makes parental permission, priestly rites, public declarations and witnesses entirely dispensable features of a legitimate marriage. Nor would they fail to challenge theological assumptions with literal forms of legal interpretation when arguing that (the waning of) consent might provide a ready pretext for repudiation of a spouse over time as was the case in Roman law.[37]

These are hardly spurious objections, and one can argue that the imposition of the consent doctrine was to some extent plagued by the law of unintended consequences. In important instances the theoretical imperatives judged to be consistent with the operations of a transcendent providence or spiritual ethic (*fas*) simply did not translate immediately into either the most practical or most socially acceptable prescriptions (*lex*). Indeed, according to the logic of consent, all social forms or agents of authority were ultimately subordinated (with regard to marital choice) to the will or grace of God as it manifested itself in relation to the individual soul and the willful disposition of the soul in marriage. At least theoretically, therefore, the doctrine of marital consent would provide much greater toleration for elopements and for clandestine marriages—acts generally regarded with moral reproach. As a result, legitimate marriage ties were evaluated according to the laws of divine dispensation in which the logic of sacramental theory and the spiritual dynamics of individual salvation superseded the customary, autocratic and often objectively verifiable criteria used in the adjudication of common law disputes. In this respect, the consensual view implicitly removed marriage formation from the domain of *lex* (which medieval thinkers associated with the arbitrary precedents imposed by secular mores and secular custom (cf. Le Bras et. al. 358ff.) and placed it within the sphere of transcendent and providential laws to be elucidated in theological terms and presided over by ecclesiastical authorities. In this context, it is easy to appreciate the twelfth-century gloss of the conflict between Turnus and Æneas depicted in the *Roman d'Enéas* as the clash of two distinct realms of matrimonial prerogative—the first based on a divine revelation and recorded in the prophecy received by King Latinus, the second based on customary rights of aristocratic power, authority, and rank as they are articulated in the protests of Turnus and in the rather compelling and original speeches attributed to the queen.

We should reiterate, however, that these apparently tolerant or "progressive" reforms received varying degrees of approbation (or disapprobation) and with few or no exceptions remained rooted in rather explicit ethical and theological concerns. Gratian, for example, emphasizes the necessity of parental choice and permission and was likely to have recognized that Roman law never employed the crucial notion of consent to prevent parental decision-making (Sheehan 14–15; Noonan, *Power* 426). In developing a contemporary definition of marital consent, Gratian was, first and foremost, drawing on canons relating to cases of egregious marital coercion and reacting against excessive forms of paternal and feudal control over betrothals (Noonan, *Power* 426–7). Although perceived, the logical ambiguities were never fully resolved. Ivo of Chartres (eleventh century) condemns clandestine marriages in terms of social "conduct" but never makes a final determination on the legal status or validity of a marriage bond contracted privately between consenting adults (Colish 2: 638). One of Gratian's early commentators speaks out against such a practice but then seems forced to acknowledge that clandestineness does not, in and of itself, prejudice what is licit and indissoluble in a proper marital bond (*Summa* of Paucapalea cited in Colish 2: 634–35 and 635, note 420). Hugh of St. Victor and Peter Lombard favor the presence of public rites and witnesses at marriages, but they too recognize that according to the logic of the consent doctrine clandestine marriages must be recognized as valid if no prohibitions stand in the way (Colish 2: 645 and 654).

These moral ambiguities also had consequences for attitudes toward adultery. As we should expect, adultery receives no explicit expression of approval from canonists or theologians, but its sinfulness comes to be defined not according to the language of concupiscence and sensual desire, not as a temptation prompted by demons, but in terms of an ethical and social transgression, in terms of an externally imposed betrothal. When Master Roland (Pope Alexander III) evokes the need to distinguish between legitimate and illegitimate forms of sexual union, between marriage and mere fornication, he does not dwell on questions of lust and sinful desire. Instead, he insists upon the concept of spousal autonomy and spousal consent and makes these the pivotal features of effective legitimacy:

> matrimonium est viri et mulieris legitima coniunctio individuam vitae consuetudinem retinens i.e. exigens, ut nulli videlicit eorum altero invito liceat continere. (Thaner 114)
>
> [*Matrimony is the lawful union of man and woman with each partner retaining his or her individual habits and customs, i.e., requiring that neither of the two be permitted to constrain the other involuntarily.*]

Here, of course, we find a clear echo of the opposition to coercion that initially motivates Gratian's doctrine. At same time, however, the logic of consent gives new importance to the very act of commitment, to the *vow* expressing the sincere *will* of two individuals above and against all formal rites and traditional conventions (Thaner 116: "Item si mulier et ante nuptias se voto alicuius abstinentiae astrinxerit, cum postea nupserit, a viro suo poterit immutari, unde Augustinus ait . . . "). In his own formulation of consent theory, Gratian refers to the decree of Pope Urban II that "Those one in body ought to be one in spirit . . . lest when, against the command of the Lord and the Apostle, the woman will have been unwillingly united to some man, she incur the guilt of divorce or the crime of fornication" (cited in Sheehan 11–12). As Sheehan observes, Gratian's own gloss "went on to say that where this occurred, the sin would redound to those who had brought about the marriage of the woman, and ended with the remark that the ruling also applied to men" (Sheehan 11–12).

Despite its growing acceptance, therefore, the consent doctrine approved by the "French School" contained fuel for numerous twelfth-century controversies in addition to those resulting from the larger *copula* vs. *consensus* debate. Most important, however, is the growing tendency among moral thinkers to credit subjective states of mind over and above objective formalities and customary conventions. For the institution of marriage itself necessarily involved (theological speculation aside) the intersection of fundamental social, legal, ethical, and individual prerogatives. As such, the important twelfth-century innovations in marriage doctrine provided a critical nexus for testing other innovative forms of moral thinking emphasizing subjective intention. In the area of marriage formation, the problematic nature of such ethical theory would come to the fore given the important role played by amorous desire—an affect that Christian thinkers had for so long associated with sin.

In any event, the *objective* and *legal* criteria sanctioning marital bonds were gradually devalued and greater significance was given to the affective disposition of the partners. Marriage was effectively being transformed from a legal and conventional institution to a more fundamentally spiritual and sacramental one. As this transformation took place, the theoretical definition of marital status became more ambiguous; authentic marriages were contrasted to inauthentic marriages, "perfected" (i.e., consummated) ones to merely initiated ones.[38] Hence emphasis was shifted away from apparent and formal legalities to the affective bonds and attractions linking individuals and spouses.

Roman law provided a fundamental concept for these vague categories of subjective sentimentality. It was the concept denoted as "marital affection" (*maritalis affectio*),[39] a notion tied directly to the concept of marital consent (Noonan, *Power* 425). The term provided canonists with a basis for attaching a moral responsibility and ethical engagement to the marriage vow. Thus while Duby's statement that twelfth-century marriage was based on a sentiment "we would call *love* today" accurately bespeaks something of the radical changes in attitude at hand, it distorts the kind of definitions that appear to have been immediately present in the mind of both Roman scholars and their medieval heirs.[40] "This idea," observes Colish, "which goes back to Roman law, means neither . . . romantic or erotic love. . . . Rather, marital affection involves respect, honor, moral regard" (Colish 2: 634).[41] Noonan points out that "neither the classical jurists nor the Byzantine codifiers attempted to give explicit, specific content to the concept. Beyond the emotion-infused intent to take the other as spouse nothing can be dogmatically attributed to affection" ("Marital Affection" 489). Peter Lombard adopts the Roman law notion but provides no meaningful elaboration. His definition of marital commitment makes reference to the mutual association of the partners (*coniugalis societatis*) guided by marital affection (*coniugalis affectio*). Nothing akin to erotic or romantic love and desire is described (Colish 2: 655).

One can only suppose, however, that such cautious imprecision would inevitably leave a void inviting speculation inspired by a broader range of popular sentimentality involving less discreet forms of secular love. Even Gratian's formulations clashed with a basic tenet of Augustinian moral doctrine, namely that only the intent to procreate should motivate marital intercourse. Instead, Gratian (relying on alternate statements in Augustine which evoke Paul's vision of marriage as a remedy for uncontrollable desire) finds in the sentiment of marital affection a positive basis for conjugal life:

> Sic contrario datur intelligi de his qui coniugali affectu sibi copulantur, quod etsi non causa procreandorum filiorum, sed explendae libidinis conveniunt, non ideo fornicarii, sed coniuges appellantur. (Noonan, "Marital Affection" 494 and note 47)

[*Thus the contrary is to be understood with regard to those who copulate
out of conjugal affection. Even if this is not done for the purpose of engen-
dering offspring but for satisfying sexual desires; nonetheless, they are not
to be called fornicators, but spouses.*]

Not only does Gratian's formulation make it clear, as Noonan observes,
that the concept of marital affection did not include the intent to have offspring ("Marital Affection" 494), it also reminds us that marriage was
coming to be defined in more and more subjective and mystical terms as a
primary manifestation of the sacramental bond between God and his secular subjects. Impressed, it would seem, by Pope Leo's declaration (against
Roman tolerance for remarriage) that the marriage bond represents an
affection that originates in God (*affectum ex Deo initum*), Gratian adopted in Noonan's words "a high view of affection," such that it became the
pivotal criterion in the formation of valid marital bonds. This view of marital affection supported the assertion that the conjugal bond was an indissoluble bond (*unde datur intelligi, quod qualiscumque fuerit vir et uxor, ex
quo coniugali affectu sibi adhaeserint, ulterius ab invicem discedere non
valent*) that superseded other more "objective" legal criteria for defining
legal marital status. In particular, it legitimized clandestine or unregistered
marriages by making marital affection (in and of itself) a sufficient basis for
legitimate marriage (*Qui contemptis omnino illis solemnitatibus solo affectu aliquam sibi in coniugem copulant*).

Gratian's teaching shows the extent to which twelfth-century marriage
reforms emphasized the freedom of the individual to choose a marriage
partner while investing that choice with sanctions anchored in both natural and divine laws. This freedom led to new visions of marriage and love
in relation to both the individual and larger Christian society as a whole. It
challenged long-standing social customs placing limitations (based on economic class, social rank or ethnicity) on the choice of marriage partners.
The language of the New Testament provided the crucial, if scant, precedents.[42] Theoretically, this position undermined the pragmatic rationales
governing aristocratic marriage strategies by emphasizing a subjective and
affective criterion essentially transcendent in nature and whose operations
were part of a divine purpose superceding human investigation and human
litigation.

Nor would it be difficult for speculators to blur further the lines of distinction between a vague canonical notion of devoted spousal affection and
expressions of amorous passion generally speaking. Here the preoccupation in secular circles with love and courtship would inevitably come into
play. Even a theologian is compelled to ask whether the union of two individuals into one is more true for spouses or for lovers.[43] Here the terms of
the speculation are defined by a common patristic topos which draws on
the Adam and Eve story in Genesis in order to define marriage as a divine
and mystical union between individuals (who are "one in the flesh") that

supersedes all other familial bonds of duty and affection. It is evident, however, that this old and familiar patristic gloss of Genesis would readily converge with a broader concept of secular love involving adulterous courtship and desire.

We have seen, from a historical perspective, that marriage doctrine reforms initiated in the eleventh century and galvanized in canon law and theological tracts of the mid-twelfth century involved important reflections on the nature of the "affection" referred to in Roman sources for marriage law. These developments are largely circumscribed by specific legal, political and theological concerns, yet we have also seen that explicit Roman and patristic authorities offered often unsatisfactory, insufficient or vague precedents that opened the door to inventive forms of synthesis and speculation and that the concept of "marital affection," while inherently subjective to begin with, did indeed remain singularly ambiguous. It seems inevitable that such insufficiencies would lead writers to look for additional "authorities" or "models" for expanding their understanding of such a pivotal concept just as parallel developments in secular lyric and larger references to pagan "history" and myths would seem destined to impinge, more or less consciously, on the minds of twelfth-century ecclesiastical authors. In other words, the new emphasis placed on the subjective and affective aspects of the marital bond and the interest in a sentiment as universal as "love" provided an immediate bridge between secular and patristic traditions, between erudite speculations on marriage and an array of old and new secular traditions and myths—a bridge between a sacred concept of *eros* that is pervasive in Indo-European mythology and Western lyricism, on the one hand, and emerging Christian meditations on the sacramental, mystical, and institutional value of marriage within the larger designs of divine providence, on the other. As such, marriage itself became for Christians what it had been for Romans, a sacred institution serving to integrate aristocratic society to the divine will and structuring the life of the secular individual in conformity with a higher law of spiritual truth. Accordingly, marriage fictions would serve as a field for broader speculation with regard to the theoretical and conceptual ambiguities of emerging doctrines and also as a means of projecting an ideal vision of aristocratic life in which marriage was perceived as playing a defining role in the life of lay individuals—individuals belonging to secular society, the third estate of the universal *Civitas Dei* alongside the saints and those devoted to religious and monastic rule. It is to these questions that we will turn in the following chapter.

In what ways do Old French marriage fictions reflect both the orthodox and the problematic tenets of emerging marriage doctrine? To what extent do the narratives serve to define and resolve the ethical ambiguities surrounding the consent doctrine and the subjective forms of affection presented as the basis for legitimate marriage formation? In what ways do larger narratives depict the divine significance of marital bonds with regard to individual experience in its sentimental, social, and spiritual dimensions? Finally, what role is assigned to marriage in relation to other bonds and institutions governing social cohesion or social conflict in feudal society?

NOTES

1. Adler ("*Militia et Amor*"): "Benoit seems to have been intrigued by the dilemma [inherent in the forces of love and war], but also embarrassed. In the end, he softly, as if in a whisper, seems to suggest a solution . . ." (15) *Militia* and *Amor* [are] both dissolved in the spirit of fraternal *Amicitia*. . . . *Militia* as well as *Amor* reduced to absurdity, leave a place for *Amicitia* of a brotherly kind" (28).

2. I will argue below that the two are clearly joined by a solemn marital rite after the abduction.

3. We should recall that Hesione is a member of the immediate royal family of Illion. This point becomes explicit later in Benoit's narrative, and we will touch on it briefly below. The censure cast upon Telamon's conduct (and we still have to consider in more detail the basis and nature of Benoit's "judgment") is, both in its moral and historical aspects, prefigured and reinforced by an earlier narrative sequence: the work's opening narration of the tale of Jason and Medea. This episode has been discussed in conjunction with my analysis of the *Enéas* in chapter 2.

4. Dictys, Book Six, pt. 3. In Benoit, Strophius appears under the name Focensis. For the corresponding episode in Benoit, see vv. 28321–28343. ("Focensis aveit non li sire / De la cité e de l'empire. / Cist haï de mort Egiston, / Si vos en dirai l'acheison. / Une fille cui il aveit, / Que de mout grant beauté esteit, / . . . Li ot donee en marriage: / Guerpie l'aveit e laissiee , / Por Clitemestran reneiee, / Que son seignor aveit mordri. / A Orestès dist e ofri / Que il ireit a la venjance, / Quar en son cuer a grant pesance / De la honte qu'il li a fait" vv. 28329–28343).

5. "La fille Menelaus cuidot, / Ço li ert vis, ço li semblot, / Qu'el n'aveit ne fine amor / Ne verai cuer de son seignor: / En la femme Hector ert sa cure; / Celi amot a desmesure" (vv. 29623–29628).

6. "[Briseida] aprent or sovent noveles: / Mout s'en rïent les dameiselles, / Mout la heent, grant mal li vuelent; / Ne l'aiment pas tant come els suelent. / Honte lor a a totes fait: / Por ço li sera mais retrait" (vv. 20677–82). As for the "dynamic" aspect of Briseida's character, see her expression of repentance and renewal (vv. 20320 ff.)

7. The irony of the fact that kinship ties link Greek and Trojan adversaries is dramatically underscored in a passage from Dares taken up by Benoit: in this passage Hector meets Telamon Ajax on the battlefield and realizes Ajax is his nephew by Hesione (v. 10131). This deters Hector temporarily from his immediate plan of setting fire to the Greek ships. The ironic contradiction between the fact that family patriotism serves as a common incentive for further conflict (wars of revenge) while Greeks and Trojan figures are often united by blood ties is another of many virtual leitmotifs elaborated by Benoit. The universal qualities of amorous passion also serve to reveal the common ground of shared subjectivity: Achilles, struck by the beauty of a Trojan woman inside the Trojan camp, is moved by love to denounce war; Troilus refers to Diomede as sharing an identical fate by virtue of their common experiences with regard to the unfaithful Briseida.

8. In the *Aeneid* , for example, Æneas says to Ascanius: "disce, puer, virtutem ex me verumque laborem, / fortunam ex aliis. . . . tu facito, mox cum matura adoleverit aetas, / sis memor et te animo repetentem exempla tuorum / et pater Æneas et avunculus excitet Hector" (XII; 435–440).

9. I have discussed Medea and will discuss Helen and Paris as well as Pyrrhus and Hermione further below.

10. For earlier references to Antenor's consummate diplomatic skill, see vv. 3393; 3250–3253.

11. For Peter Lombard, for example, beauty incites love which leads in turn to marriage (Bender 178).

12. A perspective equally consistent with regard to Achilles's fate in the *Roman de Troie*: Achilles is forced to turn his passion for a prospective Trojan bride inward and to hide his love. His ostracism leads to tragic results.

13. Among the Greeks, Nestor (in discussions with the Trojan emissary Antenor) gives voice to the same attitude: "Vos genz ne les noz ne sont une, / N'ensemble n'ont nule commune" (vv. 3479, and cf. vv. 18396–7).

14. Approved not only by Priam—as in Dares's account—but also by the Trojan princes and populace ("Heleine fu mout honoree / E mout joïe e mout amee / Del rei Priant e de sa femme / E de toz les autres del regne" vv. 4877–80).

15. In the end, Benoit leaves little doubt about Helen's love for Paris and her ability to dissimulate; see the prelude to her abduction; vv. 4319–28; 4343–54; 4371–2.

16. Of course the distinction is slight, as the notions of assent, complicity, and approval have close semantic meanings. Generally, the term *consence* has the meaning "permission," "approval," or "consent" (*Tobler-Lommatzsch* 731). The adverbial phrase *par consence* could be used to indicate specifically "willingly" as opposed to "by force" (*a force*) (*Tobler-Lommatzsch* 731; ll. 32–34). The *Tobler-Lommatzsch* also makes reference to a use of the term *consence* to indicate assent to a criminal action (731; ll. 49–50).

17. The principal features of emerging marriage doctrine are discussed in greater detail below, in the second part of the present chapter.

18. "While hunting in the woods on Mount Ida, [Paris] had fallen asleep and dreamt as follows: Mercury brought Juno, Venus, and Minerva to him to judge their beauty. Then Venus promised, if he judge her most beautiful, to give him in marriage whoever was deemed the loveliest woman in Greece. Thus, finally, on hearing Venus's promise, he judged her most beautiful." This is Dares's version (Frazer 138–139; Ehrhart 32–33).

19. This was more consistent with his status as a noble prince and royal son. We will see, however, that Benoit's allusions to Celtic motifs reintroduces certain pastoral elements. In chapter four, I will suggest that Jean Renart also aligns references to Paris with his own pastoral motifs.

20. Holy men are afoot in every land "as thikke as motes in the sonne-beem . . .." Now, "Wommen may go saufly up and doun. / In every bussh or under every tree / Ther is noon oother incubus but he, / And he ne wol doon hem but dishonour" (vv. 857–81).

21. Although detailed discussion of this fact would require a lengthy digression, it has important ramifications with regard to later developments. From Genesis to Paul, biblical scripture provides relatively little authority with regard to explicit teaching on the place of marriage in Christian life. In the time of the Church Fathers, reference to the union of Adam and Eve, to the Miracle at Cana, and to

some select lines of Matthew (e.g., XIX, 6 and 10–12; V 31–2) and Paul (e.g., Romans VII, 1–3 and 12; 1 Corinthians VII, 2–15) became the standard departure points for canonical pronouncements (see Godefroy, pp. 2057–66). Underlying later controversies was the intractable ambiguity that characterizes Paul's admonition to the Corinthians (1 Corinthians VII, 8–9): *Dico autem non nuptis, et viduis: bonum ist illis si sic permaneant, sicut et ego. Quod si non se continent, nubant. Melius est enim nubere, quam uri* ["To the unmarried and the widows I say that it is well for them to remain unmarried as I am. But if they cannot contain themselves, let them marry, for it is better to marry than to burn"]. This meant that the Church would seek to impose doctrinal prescriptions in an area where both clergy and secular lords cherished their autonomy, and that it would have to do so without reliance on any unequivocal scriptural dogma. This lack of authority would not only allow for exceptional speculation and controversy in a most contentious area of moral legislation, but deserves further emphasis given its conspicuous presence in the medieval cultural consciousness (e.g., the beginning of the "Prologue to the Wife of Bath's Tale" [vv 9–70] in Chaucer's *Canterbury Tales*).

22. Cf. Godefroy, p. 2068.

23. "Sanctitatem sine nuptiarum damnatione novimus et sectamur et praeferimus, non ut malo bonum sed ut bono melius" Tertullian *Adv. Marcion* I. XXIX [*P. L.* II, 280], cited in Godefroy, p. 2087.

24. See note 21 above.

25. Pendant les premiers dix siècles, la grande oeuvre, et combien nécessaire, de l'Eglise, a été d'introduire la moralité dans le mariage, de préciser des règles . . . non point d'approfondir le dogme. Les conséquences de la doctrine importaient plus que les formules savantes (Le Bras 2125–6).

26. Gaudemet, p. 57: "Le consensualisme, qui caractérisait le mariage romain, fut accepté par l'Eglise, qui voyait aussi dans le mariage la rencontre de deux volontés. Il lui eût d'ailleurs été difficile d'imposer à ses fidèles un autre type d'union. Devenus chrétiens, ses membres demeuraient des Romains. Pourquoi auraient-ils dû renoncer aux traditions ancestrales?" Spousal consent also privileged individual autonomy within the context of secular aristocratic ideology.

27. Paul used the term *mistirion mega*, rendered in the Vulgate Bible as *sacramentum magnum* (Godefroy 2066–70).

28. By the twelfth century "theologians and canonists generally agree that the only grounds for separation is infidelity and that if a spouse is repudiated on this account, the marriage remains in effect and neither party can remarry" (Colish 2: 662).

29. Diane Owen Hughes, for example, argues that in the sphere of secular custom, the transition from brideprice to *morgengabe* involved the transition to a consensual arrangement that privileged the reciprocal will and relation of the spouses and the shift away from family control over an economically structured marital institution ("From Brideprice to Dowry," *Journal of Family History* 3 [1978]: 262–96).

30. Gaudemet points out that the early Church Fathers clung to the Roman concept of consensualism in their rejection of any radically individualistic notion of consent in contracting legitimate marriage ties: "Tenant compte de la gravité de l'acte et de ses aspects religieux, elle [l'Eglise] s'efforce d'en garantir l'authenticité par une publicité et l'entoure de rites religieux" (59).

31. "La principale difficulté qu'il importait de résoudre, c'était donc la *fixation du moment précis* où se forme le lien matrimonial" (Ch. V. Langlois, *La Vie en France*

au Moyen Age 1924, p. 2157). "Déterminer l'instant où se forme le mariage: tel sera l'un des plus graves soucis de l'Eglise au XIIe siècle" (Le Bras 2127).

32. Citations taken from Gaudemet, p. 58 and p. 58 note 47.

33. A master of the School of Laon gives this formulation: "Ubi non est consensus utriusque non est coniugium" (Colish 639). A disciple of Abelard states: "Et hoc federatio ad primum facit coniugum" (Colish 641), and this formulation echoes Ambrose's (cf. Gaudemet, note 47, p. 58). Hugh of St. Victor states: "Matrimonium non facit coitus, sed consensus" (Colish 643). Peter Lombard asserts that the union of bodies must be accompanied by a *consensus animorum* (Colish 651).

34. "Le point qui a plus constamment occupé canonistes et théologiens, entre l'an mille et l'année 1140, c'est la formation du lien . . . . Des solennités, on ne s'occupe guère . . . . La grande affaire, c'est de déterminer la part de la volonté et celle de la *copula carnalis*" (Le Bras 2147).

35. This opposition paralleled and evoked a more all-encompassing opposition between mortal and divine laws, between *lex* and *fas* (Le Bras et. al. 358–89).) Gratian's position actually represents an attempt at reconciliation based on scholastic hair-splitting. Gratian and certain disciples from the opposing (French) school take the position that consent initiates the legitimate marriage bond while this bond is then "perfected" by the physical union or *copula* (Le Bras. 2157).

36. A contradiction ridiculed by Peter Damian, *De tempore celebrandi nuptias* (P. L. CXLV, col. 659–665: see Le Bras 2132–3).

37. On the two schools of thought see Colish 2: 630 and Le Bras 2148–51.

38. Of course this kind of dialectic informs marriage conflicts in the Old French *Tristan* and in works by Chrétien de Troyes. In an article entitled "Faux mariage et vrai mariage dans les romans de Chrétien de Troyes," for example, Maurice Accarie has used this approach to surmount moral ambiguities in Chrétien de Troyes (see, *Annales de la Faculté des Lettres et Sciences Humaines de Nice* 35 [1979]: 25–35).

39. See Noonan, "Marital Affection."

40. Noonan observes that in Roman law the term initially meant nothing more than "will" or conscious "intent," although it later took on an emotional aspect suggesting an inclination or fondness for something. He concludes that the term contrasts with the concepts of lust and sexual promiscuity and that "*affectio*, then, could suggest to a reader of Justinian "affection" in its English sense ("Marital Affection" 487–9).

41. Noonan also observes that in both Roman law and in Gratian's *Decretum*, there appears to be an implicit distinction between [marital] affection and something akin to "adulterous affection" involving an attraction which is purely sexual ("Marital Affection" 495).

42. See Noonan, "Marital Affection" 492: "Gratian quoted St. Paul "In Christ Jesus there is neither Jew nor Greek nor slave nor free man" (*Gal.* 4. 28); and where the Apostle had said of a widow, "Let her marry whom she will, only in the Lord" (I *Cor.*, 7. 39), Gratian expanded the meaning to be "Who wishes to marry, let him marry in the Lord."

43. Eudes d'Ourscamp (+1171), *Quaestiones* (ed. Pitra, *Analecta* t. 2, p. 97); cf. Le Bras 2157: "Ordre de problèmes dont les conséquences pratiques ne furent pas négligeables et dont nous aurions tort de méconnaître la place qu'il tenait dans la speculation médiévale".

CHAPTER FOUR

Courtly Narratives, Christian Sacrament:
Consent Doctrine and Social Ideology in the Old French *Roman*

Our study of Old French marriage fictions began with the observation that marriage plots were a predominant feature of the early French *roman*. We also concluded that the treatment of such related themes as love and adultery could not generally be accounted for by reference to either contemporary lyrical conventions or orthodox moral views condemning adultery and *cupiditas*. It also seems unlikely that any positive ethic of adultery was ever widely espoused whether as a social phenomenon or as a dominant poetic convention. Paden, for example, went so far as to assert unequivocally that *Le Chevalier de la Charrette*, Chrétien de Troyes's version of Lancelot's adulterous transgressions, could not have been inspired by troubadour motifs because adulterous love was not truly a central tenet of troubadour lyric. Paden's conclusion supports our assertion that secular marriage plots cannot be reduced to celebrations of or reactions against a supposed ethic of adulterous love just as the elements making up the plots do not truly conform to familiar lyric conventions. At the same time, however, it seems almost impossible to make sense of Chrétien's *Lancelot* without reference to lyric conventions. It should be obvious that the adulterous plot (accounted for by Celtic sources) need in no way exclude lyric motifs. Nor does this fusion of lyric motifs with the triangular structure of the Celtic sovereignty rivalry need imply that *fin'amor* is *a priori* adulterous.[1]

Indeed, contemporary marriage doctrine controversies can explain the primary motivation for and interest in adulterous transgression in the first place. In virtually all of the works in our corpus, a consensual model of marriage is presented as a superior alternative to an essentially coercive marriage practice based on feudal customs. From a typological perspective, the clandestine love affair—which is regularly inflected with an aura of providential destiny—reveals the inadequate and tainted nature of the coercive bond from which the adulterous or clandestine affair inevitably derives.

However, Church morality and patristic culture as such do not provide abundant scriptural, historical, or social precedents for this view of marital mores. This gap between patristic precedent and emerging marriage doctrine can be made quite clear when we recall that the same eleventh- and twelfth-century marriage reforms that sought to affirm the sacramental value of love and marriage in the temporal affairs of secular subjects also insisted on enforcing the traditional moral prohibitions against marriage and concubinage with regard to the religious orders. In order to discuss the representation of both amorous and marital mores in twelfth-century narratives, one must bear in mind that a new sacrament is now emerging—a sacrament sanctioned by Church authority but without precedent or parallels with regard to the religious order strictly speaking. For obviously, matrimony was a state proper only to individuals outside the clergy and who had not taken monastic vows. Indeed, just as celibacy became a principal virtue of the religious subject, the conjugal union—with its virtuous affections and deeper providential aspects—served (according to the new doctrine) as the defining sacrament in the life of the secular Christian lords who were the primary agents in the realm, not of spiritual, but of historical (national and political) providence.

The emergence of such a radical shift, one in which "moral teaching" as such could no longer be articulated in absolute and universal terms but rather in terms of relative social orders (one clerical and monastic, the other temporal) obviously has profound implications. This development suggests what our previous survey demonstrated—that neither reference to lyric tradition nor reference to a clerical irony based on patristic morality can account for narrative representations of love and marriage in the twelfth-century *roman*. In fact, such a shift implies an inherent impasse between discrete spheres of social activity and discourse, thereby transforming the traditional relationship between the secular subject and the pastoral guide, between secular experience and conventional forms of moral teaching. The requirements of love, courtship, and marriage will come to define the providential and "sacred" itinerary of the *secular* subject.

Our immediate appreciation of Old French marriage fictions rests initially, therefore, on our understanding of specific doctrinal developments, especially the principles of marital consent and marital affection at the center of emerging canonical and theological doctrines. In all of the later works in our corpus, marriage fictions are also portrayed according to the norms of marital consent. Arranged marriages (devoid of amorous and marital affections) end in misfortune while the fatalistic prerogatives of love prove virtually invincible. It is also true, as in Benoit's treatment of the abduction of Helen, that illicit or clandestine lovers are in some way vindicated—as is the case in the Old French *Tristan* narratives (Béroul and Thomas), in *Partonopeus de Blois*, *Girart de Roussillon*, Chrétien's *Cligès*,

Gautier's *Eracle*, and Jean Renart's *Escoufle* and *Guillaume de Dole*. In fact, the emergence of the *roman* as an Old French genre not only coincides with the emergence of new marriage doctrine but seems intimately tied to it. Once we have surveyed the persistent parallels between a social and moral doctrine on the one hand and literary themes and representation on the other, however, we must attempt to define a number of more subtle and more complex typologies. Indeed, our analyses of the *Roman d'Enéas* and the *Roman de Troie* have already shown that the marriage motif is both influenced by emerging doctrine and inseparable from a much broader typological vision of historical providence. The new "space" or field of moral value and discourse inherent to the secular orientation of the marriage sacrament has already been seen, in the two *romans d'antiquité*, to invite an elaborate synthesis of temporal and Christian concerns that draws upon mythological, epic, and lyric traditions.

MARITAL CONSENT IN OLD FRENCH MARRIAGE FICTIONS

> *J'ai amors a ma volente . . .*
> *J'ai amors a ma volente*
> *tels com je voel*
> —*Guillaume de Dole* (vv. 5440–43)

In virtually all of the works in our corpus, coercive marriage practices motivated by competing political and economic interests come into conflict with consensual bonds motivated by mutual love.[2] At the end of the *Roman d'Enéas*, the author of that work uses the marital rivalry between Eneas and Turnus as a vehicle for contrasting two competing models of marital rights and destiny.[3] Turnus and his mother seek to protect their ancestral and territorial status by suing for Turnus's marriage to Lavinia, and Turnus asserts his right to Lavine's hand as a function of his aristocratic rank, wealth, and power. Turnus's mother likewise condemns the unfounded pretensions of the disenfranchised and foreign knight Eneas. It is interesting that the author of the *Enéas* goes so far in undermining the Trojan's own claim on Lavine. In doing so, however, he makes it clear that his larger design is to introduce into the narrative an alien and competing model of marriage deriving from a mystical order of prophetic revelation and aligned with the manifest destiny of a sweeping *translatio imperii* and absolute sovereign dominion.

In creating this bold vision of the secular order and its historical unfolding, the author of the *Enéas* also sought to fill in the logical gap left by his critique of conventional secular marriage customs. In other words, how did the secular subject—typically circumscribed by the laws and customs of his own social community—experience his own marital destiny outside these conventional realities? Here, Ovidian sentimentality (divorced from its often deceitful or cynical ethos) provided a necessary link in the evolving synthesis.[4] The powerful psychology of love, the experience of "falling in love," the undeniable truth inherent in the Ovidian credo *omnia vincit amor*, all of these Ovidian elements provided literary models for representing a marital order based on affection. The undeniable "experience" of love as a transcendent passion reflected (on the microcosmic and private level) the workings of a providential (macrocosmic) order whose prerogatives transcended social custom in the same way that amorous desire is commonly portrayed as an impetus to acts of individual rebellion or transgression. The socially transgressive aspects of erotic love have, in fact, only a contingent status (as the outcome of conflicting marital mores), but serve nonetheless to dramatize potential ethical and social conflicts resolved only by reference to an implicit hierarchy of ethical values.[5]

We have also seen that the relation between temporal love and providential history is made quite explicit in Benoit's critique of marital mores in the *Roman de Troie*.[6] In this work the tragic aspects of the ancient Greek world are glossed in terms of marital disorders, and the doctrine of marital consent positively reorients the seminal topos of adulterous transgression contained in the famous tale of Paris's abduction of Helen. By contrast, the interwoven tales of clandestine love and marital discord provided a further commentary on the importance of legitimate consensual marriage bonds. In this text, it is the departure of Crisede for the Greek camp that elicits problems of popular censure and higher ethical judgment. What about subsequent Old French narratives? To what extent do the representations of love, courtship, marriage, and adultery in our larger corpus of works consistently reflect the principle of spousal consent that serves as the cornerstone of emerging ecclesiastical marriage doctrine and reform? To what extent is the "problem" of adultery resolved or attenuated by the critique of coercive customs and simplistic moral judgment?

In some works, as for example the two *romans d'antiquité* which we have examined in detail, the essential tenets of the consent doctrine serve as a means of orienting and glossing existing narrative traditions. The marital conflict culminating in the fall of Troy provided a dramatic and seminal *exemplum* for illustrating the profound consequences of marital norms. It also served to demonstrate what Church doctrine now professed, not only that consent and affection form the basis of an enduring and properly consummated union, but that the institution of marriage itself (with its sacramental dimensions) governed the progress of private and political affairs. Finally, one can speculate that the important role of marriage with regard to the secular orders—as it was so defined by emerging Church doctrines—provided a singular impetus (alongside the interest in dynastic origins) for the resurrection of pagan heroes such as Paris and Eneas, that is for heroes whose epic deeds were intimately tied to marriage dramas. These perspectives might also account for the twelfth-century interest in the figures of Tristan and Iseut. We saw, for example, that the illicit affair between Tristan and Iseut does not culminate in the elaboration of the archetypal adulterous love triangle, but in a kind of marital quadrangle in which the tragic conflict is "explained" not in terms of adulterous love *per se*, but in terms of marriages devoid of mutual consent and affection. Indeed, Tristan's marriage to Iseut aux Blanches Mains does not serve to elucidate the notion that aristocratic marriages exclude amorous sentimentality—for both heroines are in love with Tristan—but demonstrates the need for mutual affection and consent as essential components of a prosperous and enduring marriage bond. But even more importantly, the tragic aspect of the Celtic drama could be exploited as a critique of aristocratic marriage customs which dissociate the institution of marriage from mutual bonds of subjective affection and consent.

In the tale of the sage and hero Eracle by Gautier d'Arras, marital fidelity also depends upon amorous volition. The fact that Gautier saw fit to interpose a tale of marital conflict and adultery into the fundamentally hagiographic outlines of his story already suggests something of the theological dimensions underlying the secular drama as it relates to marriage practice and marital conflict. As a secular lord the emperor Laïs is faced with two fundamental obligations. First, he must insure the integrity and continuity of his dynasty by means of a suitable marriage. Second, he must leave behind his wife when he is called upon to defend his realm. In his absence, of course, his wife remains unsupervised and her conduct open to suspicion. The inherent dilemma reveals the extent to which lay marriage practice is flawed because of its essentially coercive nature. The "integrity" of the relation depends upon external controls, and Laïs, like most medieval aristocrats, is an itinerant lord who must leave his court either to hunt, to wage war, or to carry out diplomatic missions. *Eracle* highlights the inevitable limits of sequestration as a guarantee against adulterous transgression.

In the Old French *Tristan*, in *Flamenca*, and in *Cligès*, the devices contrived to safeguard the chastity of the wife merely emphasize the virtually limitless inventions of those who serve *Amor*. Much of the humor in *Flamenca* derives from the parallel machinations of the jealous husband Archimbaut and the lover Guilhem. Nothing could be more impenetrable than the safeguards erected by the jealous husband, nothing more convoluted than the painstaking machinations carried out by the adulterous suitor. In *Cligès*, the sarcasm tends to redound more evenly upon both the coercive husband and the inventive lovers. But, as in the Old French *Tristan* story, the magic potion that serves to delude the husband into falsely believing he has physically consummated the marital union stands as a metaphor for another delusion, namely the aristocratic belief in a coercive marriage bond void of authentic affection. Ironically, it reveals with burlesque absurdity the inherent futility of all coercive strategies, be they customary or adulterous.

In *Eracle*, therefore, it is not surprising that Laïs's wife Athenaïs will be caught having an adulterous affair with a young suitor named Paridès.[7] Somewhat paradoxically, this blatant ethical transgression will be readily dismissed by the wise and pious Eracle. Instead, Eracle reproaches Laïs for having sequestered and mistreated Athenaïs despite previous admonitions to the contrary (vv. 2969–3220; vv. 4961–5044). Anthime Fourrier points out that Laïs's misfortunes evoke a proverbial Ovidian query as to whether jealousy or trust can better protect a husband from infidelity (Fourrier 230 and 262–3; cf. *Ars amatoria* III; 611ff.). The lesson drawn by Fourrier is consistent with our own reading:

> Or, dans les affaires du coeur, la vertu n'est point l'effet de la contrainte ni des conventions extérieures: elle naît du consentement personnel. (261)
>
> [*Now, in matters of the heart, virtue is not a consequence terms of coercion or external conventions. It arises from personal consent.*]

In *Cligès*, the initial marriage idyll—based on the love between Soredamors and Alexander and set in Arthur's court—defines marriage in terms of mutual consent and affection. The subsequent marriage drama involving Cligès and Fénice also ends in the triumph—albeit after many ironic trials and tribulations—of mutual love. In *Girart de Roussillon*, the King's attempt to impose his will on Girart and Elissent leaves him with a bride in name only. Not only do Elissent's affections remain directed toward Girart, but the two affirm their own bond by virtue of a mutual vow of fidelity and affection. In this epic setting, the inherent inadequacy of the coercive model is mirrored by the weak feudal bonds that underlie the protracted hostilities between king and vassal. *L'Escoufle* and *Partonopeus de Blois* contain similar conflicts. In both works young lovers ideally fitted to one another and attracted by mutual love find themselves at odds with parental prerogatives and must resist larger interests that conspire to bring about their separation. The transcendent nature of the amorous desire is revealed in the power it has to triumph over external afflictions and shape both individual and political destinies. Indeed, *L'Escoufle* and *Partonopeus de Blois* make quite explicit connections between a consensual marital ideology and reliable feudal bonds (feudal bonds governed by subjective forms of loyalty, honor, and good faith). Therefore, the transcendent nature of the marriage sacrament finds its rationale not in explicitly theological terms, but according to the logic of the secular state and secular subject and in perfect accord with its uniquely secular role. This correlation between feudal and marital bonds represents more than a lyric metaphor and provides a dominant motif in other works such as Béroul's *Tristan et Iseut* and *Girart de Roussillon*.[8]

MARITAL CONSENT: *FLAMENCA*

In *Flamenca* we find elements similar to those in *Eracle*—a married nobleman crazed by jealous fear loses his wife to an adulterous rival. Although

Archimbaut becomes a burlesque character made in the image of the jealous husband of the Occitan *canso*, the author develops the question of spousal consent as a way of both foreshadowing and explaining the marital conflict. At the point where the extant text of *Flamenca* begins, Count Guy of Nemours has summoned his counselors to discuss the respective merits of Flamenca's suitors. Although the discussion initially excludes Flamenca herself, the counselors (having expressed their favor for Archimbaut) advise the Count to ask first the opinion of both the bride and her mother. Due to a substantial manuscript lacuna (following line 58 of our edition) we can, unfortunately, only surmise as to how the author treated the here crucial question of Flamenca's own will and consent in the matter of her own betrothal. Other passages suggest, however, that Flamenca's consent was lacking, and that the ensuing tale of infidelity was meant to be seen as the inevitable outcome of an imposed betrothal motivated by the kind of prerogatives that Duby emphasizes in his model of customary aristocratic marriage practices. Both of Flamenca's parents seem to make their own interests and affections a primary consideration over and above Flamenca's own will and any regard for the true merits of the rival suitors. Indeed, Archimbaut clearly has an inferior social rank in relation to the other suitors (who are foreign kings). We learn that Flamenca's parents prefer Archimbaut because they desire to keep their daughter near to their own court. Also, the purely rhetorical praise given to Archimbaut by his counselors suggests that they are simply fearful of expressing any judgment that might displease their lord. As the counselors address Count Guy, however, they also make explicit a more pragmatic motive for marrying Flamenca to a local feudal prince such as Archimbaut, namely the opportunity to forge a convenient military alliance:

> Lo cors nos o dis e amors,
> Mais vos faría de socors
> En Archimbautz, s'ops vos avía
> Que-l reis Esclaus ni-l reis d'Ongría (vv. 33–6)

> [*Our hearts and the love and good faith we owe you tell us that Lord Archimbaut would be of greater aid should you have need than would be the Slavonic king or the king of Hungary.*]

It is in fact (and somewhat ironically) made clear that customary assumptions exclude any concern for Flamenca's own willful interest and consent. Of immediate concern are Count Guy's personal and political interests. The Count relishes the chance to fix an alliance with the Bourbon court ("Si Dieus mi dona / Un'aventura que m'es bona / Non sabra bon a totz ensems? / Ieu ai desirat mout lonc temps / C'ap N'Archimbaut agues paría" [vv. 3–7]). Likewise, it is the Count's consent (not Flamenca's) and the intractable formalities of aristocratic etiquette that will determine the course of events:

Per son anel Domini-m manda
Que Flamenca penra si-m voil;
Mout i faría gran ergueil
S'ieu d'aizo dizía de no. (vv. 10–13)[9]

[*With his seigniorial seal he has sent to inform me that he will take Flamenca if I consent. I would appear very arrogant were I to say no in this matter.*]

The Count's use of the word *aventura* (in line four of the text previously cited) alongside the reference here to Archimbaut's seigniorial "ring" (*anel*) says something as well, I believe, of an intended irony. The lexical connotations of the word *aventura* alongside the reference to the ring evoke a competing model of marriage based on amorous courtship. Ironically, these associations only call our attention to their absence within the customary feudal model while underlining Flamenca's very passive role with regard to her own marital destiny. Thus the present events adumbrate the inherent flaws of the customary aristocratic marriage and foreshadow Archimbaut's own fate (as the spurned, jealous and cuckolded husband). Of course, mutual desires will ultimately prevail over arranged betrothal, and Flamenca consents to Guilhem's advances. From this perspective, the problem of the relevant manuscript lacuna is made less significant. It seems obvious that any mention made of Flamenca's own "consent" in her betrothal to Archimbaut would be largely nominal (or purely conventional, like the counselors' praise of Count Archimbaut) and that her will in this context remains circumscribed by paternal authority and social convention.

The ironic absence of amorous motifs also introduces a larger commentary on the nature of the affective relation between Flamenca and Archimbaut. Anything akin to mutual marital affection appears to be absent from the marital union. While Archimbaut seems to burn with a mad passion, Flamenca appears only to convey a mix of passive complicity and deliberate dissimulation (vv. 286–92). The muted evocations of Flamenca's inner sentiments (e.g., vv. 344–8) become all the more meaningful when contrasted with the insistent references to Archimbaut's own burning desires (e.g., vv. 158ff.). The relatively decadent nature of Archimbaut's crass and libidinous appetite is revealed not only in the kind of calculated seduction that marks the couple's first night together but also by contrast with Guilhem, Archimbaut's future rival. Only when Guilhem enters the action does the poet make reference to *Amor* (v. 1783). In the depiction of Archimbaut's desire for Flamenca, on the other hand, the allegorical deity remains conspicuously absent. Therefore, although the adulterous courtship between Flamenca and Guilhem is in many ways a dramatic elaboration of troubadour motifs, the larger plot structure serves to contrast a customary feudal model of marriage based on coercion with an affective form of courtship involving mutual consent.

Indeed, the depiction of Guilhem as the clandestine lover does not entirely conform to lyric motifs. While in some monologues (that impart a more esoteric view of the amorous union) the notion of mutual affection converges neatly with the cultivated eroticism and pseudo-mysticism that characterize *fin'amor* (e.g., vv. 4052–6 and 4064–8), other sentiments seem to depart from familiar lyric conventions, as in the following expression of sentiment from one of Guilhem's interior monologues:

> Ma domna es e fons e rosa . . .
> Et on mais pens mai voil pensar . . .
> Ara vejas doncs que faría
> S'entra mos brasses la tenía
> Que la sentis et la baises
> Et a ma guisa la menes!
> Mais trop ai dig senes comjat
> Quar de son tener ai parlat,
> Quar non s'atain aisi la tenga;
> Non voil que per orat m'avenga
> Si non avía son autrei.   (vv. 4701, 4705, 4707–15)

> [*My lady is both fountain and rose . . . . and the more I think on her, the harder it is to stop . . . . See then what I would do if I held her in my arms, that I might feel her and kiss her and do with her as I please. But I have said too much without permission, for I spoke of holding her when it is not proper that I hold her so. I do not wish to have my prayers granted without first having her consent.*]

Like the canonists and theologians of the twelfth century who ponder the exact sentimental value of the Roman *affectio maritalis*, the author of *Flamenca* attempts consciously to define a proper form of affection that is distinct from the arrogant passion of Archimbaut and restrained in relation to lyrical hyperbole. Guilhem's deferent sentimentality contrasts sharply with the raging desire of Archimbaut. Somewhat paradoxically, the coercive model contains virtually exclusive forms of censored marital conduct: insufficient affection and volition on the one hand, and an excessive or "lecherous" passion on the other. By contrast, the love associated with *Amor*, albeit adulterous in this case, is characterized by some degree of respect, humility, and sincerity.

In all the works in our corpus, therefore, the Church's consensual theory of marriage provides the underlying logic governing the plots. At the same time however, it is in the realm of secular life that the marital order finds its proper place—in a realm of competing attitudes, customs, and prerogatives and not in the realm of rarefied theological ideas. In other words, the theological view of marriage found its own arguments within the context of the secular tensions that could disrupt marital harmony. In terms of textual strategy, the outcome of such conflicts can be said to "reveal" a transcendent, theological order.

Later descriptions of the marriage between Flamenca and Archimbaut
lend support to this view while corroborating our assertions about the significant role played by the notion of consent. In fact, the question of
Flamenca's consent is mentioned explicitly in a later passage. The relevant
exchange takes place just before the wedding day and is initiated by
Flamenca's father who, without any prior arrangements and apparently
seeking to impress Archimbaut, invites his future son-in-law to visit
Flamenca in her own chamber:

> Le cons lo pres per miei la ma,
> Ab lui vas la cambra s'en va
> Et a Flamenca lo presenta;
> Non fes semblan que fos dolenta
> Mas un pauc estet vergonosa
> Le coms di: "Vesi vostr'esposa,
> N'Archimbaut, si-us plas, prendes la."
> "Sener, si en leis non rema
> Anc ren tan volontiers non pris."
> Adonc la piucella somris,
> E dis: "Senher, ben faitz parer
> Que-m tengas en vostre poder,
> Qu'aissi-m donas leugeramen;
> Mas, pos vos platz, ieu I consen." (vv. 267–80)

> [*The count takes him by the hand and leads him to the chamber where he
> presents him to Flamenca. She does not show that she felt sad, but acted
> slightly timid. The count says, "Here is your wife, Lord Archimbaut, if it
> pleases you, take her." —"Sire, if she holds no objection, never have I
> taken anything so willingly." And then the girl smiled and said, "Dear
> sire, my father, you clearly make it appear that you hold me in your power,
> that you should hand me over so easily. But, since it pleases you so, I con-
> sent to it."*]

In this passage the poet carefully aligns and ironically juxtaposes the related notions of will and consent (*volontiers, poder, consen*). Although it is
Flamenca's marital destiny that is at stake, it is her father who governs the
events and who, as Flamenca says, seeks to flaunt the power he exercises
over her ("ben faitz parer / Que-m tengas en vostre poder"). Flamenca's
explicit "consent" (*consen*), on the other hand, has an empty and ironic
ring. In the same breath with which she gives her consent, she acknowledges her own powerlessness over her marital destiny. This implicit hesitation, with its muted and perhaps only half-conscious resentment, is evoked
in the poet's fleeting reference to the bride's demeanor ("Non fes semblan
que fos dolenta / Mas un pauc estet vergonosa") and in Flamenca's own
slanted use of the word *leugeramen*. Her sentiments are juxtaposed, ironically, with those of Archimbaut who, elated, finds the marital convention
entirely to his will and liking ("Anc ren tan volontiers non pris").

Thus, while the consensual theory derives from canon law and Roman law, the very notion of consent itself involves an act of volition which is central to the dignity and the *honor* of the aristocratic subject—free by right of birth. Church doctrine notwithstanding, *Flamenca* dramatizes the tension that results from the clash of corporate and individual interests.

MARITAL CONSENT: *PARTONOPEUS DE BLOIS*

In *Partonopeus de Blois*, the principles of mutual affection and consent are also defined against the backdrop of competing aristocratic customs and interests. The conflict between an affective (sentimental, psychological, and ennobling) model of courtship and customary aristocratic practice is ostensibly dramatized by a spatial contrast between a fairy court and a normative aristocratic court. However, the apparent spatial boundary, like the apparent alterity of Mélior's court, is quickly reduced to the status of pure psychological metaphor. We discover that both Partonopeus and Mélior suffer similar dilemmas, their marital desires in conflict with the prerogatives of the courtly interests that surround them. The marvelous aspects of Byzantium are nothing more than a metaphor for the secret and enchanted space of the endangered desires that bind the two main protagonists. Indeed, it is the ambiguous status of Mélior upon which the shifting perspectives will pivot. Is the world which Mélior inhabits incompatible with the aristocratic norms of the Frankish court? Is amorous passion a dark, subversive, and alienating curse associated with demonic fairies or is it an ennobling drive? Is the fantastic wealth associated with Mélior a dangerous, seductive lure or a metaphor for the limitless joy of a partnership forged from mutual love, desire, and consent? The anonymous author of the Old French *Partonopeus de Blois* exploits the various associations surrounding Mélior. As we illuminate this strategy, I would like to focus specifically on the way in which the lovers' happy marital conclusion (based on mutual love and consent) is contrasted with the norms of aristocratic marriage.

From the normative perspective of the Frankish court, Mélior's seductive allure poses a threat to the marital fortunes of a promising heir. When the hero's mother learns that her son is infatuated with an unknown enchantress, she immediately seeks the aide of the Frankish king and proposes to betroth Partonopeus to the king's own niece. By evoking the deeper superstitions associated with the apparently mysterious Mélior, the poet brings to the fore the dynastic and political import of marriage on the one hand, and the threat posed by subversive eroticism on the other. The anxiety expressed by Partonopeus's mother relates not only to the question of betrothal, but to the effects of demonic powers:

> Sire, fait-el, perdu avés
> Partonopeus, se nel gardés.
> Quant perdus fu en le forest,

Uns diables à soi l'a trest:
En samblant de feme se mist,
Et al dolant tos ses bons fist. (vv. 3933–38)

[*Sire, she said, you will find you have lost Partonopeus if you do not take
heed. When he got lost in the forest, a demon got hold of him and after
taking the form of a woman, treated her victim to all her charms.*]

Thus Partonopeus's *"rice aventure"* represents a threat on numerous levels.
There is the mother's fear of losing her son, both geographically and in
terms of affection ("nus del mont / De tos cels qui furent et son, / N'aiment
rien tant com mère fis" vv. 3855-7)—a fear similar to that we saw
expressed by Flamenca's father. There is the fear that the young prodigy
will also be lost to his crown and homeland—as expressed here in the
mother's strategic appeal to the king ("perdu avés . . . se nel gardés").
Finally, there is the fear that these transgressions amount to sacrilege, as
conveyed by the underlying metaphor of becoming "lost,"—lost in the forest, lost to one's family and court, a soul "lost" to the devil.

The mother's dynastic marriage designs, on the other hand, should
ensure her son's safety and "salvation" by having him married to the king's
own niece (vv. 3951-3; 3957-8). The mother also points out that this
arranged marriage will resolve the other threats as well—Partonopeus will
be formally united in this way with his homeland and the Frankish crown
and will remain under the authority of both the mother and the king. In
actuality, however, the mother's actions—her intended "reversal" of her
son's fate—becomes the pretext for a larger ironic reversal on the part of
the poet who deftly transfers the demonic fairy *tropes* over to his own critique of a fundamentally coercive aristocratic marriage practice:

Dounons-le à lui: si remanra [Partonopeus],
Et vos dirai conment c'ira.
Mandés-le en haste, c'orendroit
Sains tos essoignes à vos soit.
Dementres me faites livrer
Deux beaus bouceaus de bon vin cler;
J'atornerai l'un à mon fis,
Que s'il en boit deux trais petis
Tos ert en autre sens tornés,
Et fera bien nos volentés. (vv. 3959-68)

[*Let us give him to her: then he [Partonopeus] will continue living here
with us. I will explain how we can achieve this. Send him an urgent sum-
mons, that on the spot without excuse for any delay, he come to you.
Meanwhile, have two small kegs of good wine delivered to me. I will give
one to my son; once he has taken two small sips, he will be entirely of
another mind and will do just what we wish.*]

Once enlisted into the mother's marriage scheme, the king's niece herself
will take on aspects analogous to those typically associated with the

demonic fairy figure whose potions and charms serve to delude and "capture" an unsuspecting victim:

Vostre nièce nos servira,
Et de cel vin l'abeverra:
Nos beurons de l'autre picier,
Si lairons lui et le plaidier.
Se poons faire qu'il l'afit
Retenu l'en aurons, je quit. (vv. 3969–74)

[*Your niece will prove useful, and will serve him this wine. We will serve ourselves from the other pitcher, and then we will counsel and cajole him. Thus we can get him to promise to do it, and I believe we will get him back for good.*]

In his depiction of the arranged betrothal, therefore, the poet transfers to the aristocratic marriage the fearful and supernatural forms of seduction and deceit that have been evoked in association with the fairy figure Mélior. In the end, the fairy motif is exploited as a way of criticizing feudal marriage custom rather than affective courtship.

This strategy is even more evident when the mother and king actually implement their plot (vv. 3977ff.). In his description of the events, the poet dwells on the role played by the alluring physical charms of the king's niece in conjunction with the intoxicating effects of the wine. Now, however, the wine is referred to as a swift and violent "potion" ("Forte est la puisons et novele" v. 4004). As was the case in *Flamenca*, the poet links coercive marriage practices with aberrant forms of erotic seduction and deception.[10]

Reference is also made to the economic interests that typically motivate aristocratic marriage arrangements. Given the context, however, the king's description of his niece's substantial possessions conjures up the images of wealth that are also a feature of Mélior's fairy mystique:

En lor consel s'enbat li rois,
Et promet lui de rentes crois,
Et de casteaus et de cités,
De viles et d'onors assés. (4021–4).

[*The king rushes ahead with the negotiations, promising him increased incomes, castles, cities, and no small number of towns and other privileges.*]

The coercive and seductive devices associated with the demonic fairy figure which ostensibly threaten aristocratic prerogatives, or at least *corporate* aristocratic prerogatives, are now associated with the coercive impositions of "normative" aristocratic marriage practice.

The two logical corollaries of this ironic reorientation of value are 1) a demystification of Mélior's personage and domain, and 2) the positive orientation of consensual love in relation to larger feudal and Christian ideologies. From the perspective of competing marriage models, the poet's reversal of values is reflected in the way Mélior's own external restraints also reflect customary feudal norms. Seemingly fantastic at first, at the least ambivalent, Mélior and the realm she inhabits ultimately come to exhibit no fundamental alterity when compared to the Frankish domain.

Although we see Mélior initially from the perspective of Partonopeus's exploits, later narrative developments focus equally if not more on the matrimonial imperative and the political rivalries that surround the figure of Mélior herself as an exceptionally wealthy heiress. The same counselors and guardians who exhort her to marry will also seek to serve as judges who control her marital destiny. As is the case for Partonopeus, their judgment is based on the wealth, nobility, and status of the potential suitors and excludes from consideration the individual desires of the bride herself. Hence, Mélior's marital aspirations are also, like those of Partonopeus, beset by external threats associated with conventional aristocratic marriage customs. In fact, the competing aristocratic interests are so numerous, that the lay model itself seems to ensure its own impasse. The inherent flaws of the coercive lay model are implicitly criticized as Mélior herself tells her sister Urrake about the inevitable contention plaguing the assemblies where the king and noblemen of the realm discussed her eventual marriage:

> Si furent mi conte et mi roi
> Et tuit cil qui tienent de moi;
> Nus hom n'est de menor chasez
> Ne fust à tel conseil mandez,
> Et fu li consauz chascun jor
> De moi faire prenre seignor:
> Prisme l'empéror d'Espaigne,
> Et après celui d'Alemaigne,
> Et puis le juesne roi de France.
> Si ot entr'ax tel mesestance,
> Que por nul droit ne por nul tort
> Ne porent estre à un acort,
> En chascun ot tant à blasmer
> Qu'il ne's vorrent de nul loer. (vv. 6461–74)

[*My count and my king were there and everyone with any interest in my state. No one present was a petty landholder, nor was any such inconsequential lord even invited. Everyday the deliberations took place concerning the selection of a suitable husband for me: first the king of Spain, after him the king of Germany, next the young king of France. Between them there was such troubling disagreement that neither on account of any right nor on account of any offense were they able to come to any consensus. Each of the suitors had so many faults that they were unwilling to approve any of them.*]

It is this impasse that leads to the organization of a competitive tournament. However this solution also makes no provision for Mélior's own volition. Mélior laments this outcome, because, as her sister Urrake succinctly explains, she will be *forced* to love the knight who wins:

> De Partonopeus est nient
> Qu'il soit à cest torniement.
> N'est mais en vostre cois d'ami;
> Li vostre vos donront mari:
> Il coisiront, vos amerés;
> A lor cois vos amors donrés. (vv. 6737–42)

> [*The presence of Partonopeus at the tournament will be of no conse-
> quence. Nor does it matter whom you choose as a favorite—your court
> will choose your husband. They will choose, you will love. To whomever
> they agree upon, you will give your love.*]

In another twist of fate, however, it is the tournament which offers Mélior
the possibility of a happy resolution. Mélior is too blind from despair and
remorse to perceive this, but the tournament offers an outcome that conforms to a competing view of marriage. Ostensibly it does deny Mélior the
right to "choose" her own husband, but it does provide a basis for reconciling patriarchal authority with the values that account for Partonopeus's
privileged status (*i.e.*, as Mélior's elected suitor) in the first place. The par-
ticipant at the king's council who proposes the tournament belongs, not
insignificantly, to a lower economic status than most of the barons who are
present. Standing outside the entrenched economic interests that have made
Mélior's marriage into an issue of inevitable feudal conflict, Hernols de
Mal-Bréon—although a marginal figure (a "vavassor de petit pris"v.
6485)—seems to be no less wise ("moult est saiges de raison, / Un grant,
un vielz, un lonc chenuz" vv. 6478–79). After succinctly describing the self-
interested and contentious nature of a debate which effectively excludes
any concern for the welfare of Mélior herself (vv. 6503–6514), he highlights the flawed logic and corrupt motives underlying customary aristocratic marriages:

> Ma dame a roialmes plusors,
> Et contéez, hautes hennors,
> Et riches citez et chasteax,
> Et viles plaines et caseax,
> Et tant du sien, la Dieu merci,
> Qu'à cois puet aler de mari.
> N'a soig de plus qu'el a adès,
> Fors qu'el se tiegne bien en pès,
> Et à ce praigne cel garant
> Que tuit l'en soient bien voillant.
> S'el prent home por manentise,
> Il ert tornez à covoitise,
> Et s'en prent nului por richece,
> En dira que c'est par destrece;
> S'el prent home de halt paraige,
> Honte en aura et nos domaige. (vv. 6517–32).

[*My lady has numerous realms and territories, high honors, rich cities and castles, country lands and rustic dwellings. She has so many possessions (thanks be to God) that she can readily take a husband of her own choosing. She has no need of more than she already has, except that she enjoy true peace; with this comes the assurance that all will wish her well. If she take a man on account of his power, he will become an object of envy. Should she marry for money, it will be rumored she married out of desperation. If she takes a man of high nobility, shame and injury will result.*]

The reasoning here seems to go very directly against aristocratic conventions and is all the more remarkable to the extent that the logic itself seems strained at points. Yet it reveals with what conscious design the poet seeks to discredit customary aristocratic attitudes toward marriage. The underlying logic also gains somewhat in clarity when Hernoul goes on to imply that it is marriage for economic interest as opposed to marriage for harmony, peace, protection, and personal honor that is deplorable and not marriage to a wealthy man as such:

A ce tirons sor tote rien,
Qu'el se marit et bel et bien.
Bien le face por son profit,
Et por s'anor beau se marit;
Et si estuet que ses mariz
De bones tesches soit garniz,
Qu'il soit gentix hom et leax,
Et chevaliers et bons et beax,
Saiges et preuz et mesurable,
Et de parole soit estable.
Qui ce porroit en un trover,
Cil ne feroit à refuser;
Et vos sai bien dire conment
Il ert trovez, mien escient. (vv. 6533–46).

[*With this we should conclude above all else that she be married once and for all. Choose well for her own sake, and may she be admirably paired for the sake of her honor. And so it is necessary that her husband be graced with noble airs, be a good and handsome knight through and through, wise, valiant, humble, of firm and sensible speech. If a man with all this could be found, he should not be rejected. Well can I tell you all how we will find just such a man.*]

Of course, the means that Hernoul has in mind here is the tournament, as a test of merit. But it is the shift in underlying criteria that is most significant. We notice that feudal values have not necessarily been abandoned in exchange for an explicit ecclesiastical doctrine of marriage by consent, but that an effective *rapprochement* has taken place between the implicit ecclesiastical doctrine that sanctions Mélior's choice and an alternate set of feudal values based not on corporate economic interests and prestige as such but on chivalric refinement, honesty, and gentility. In other words, the new criteria do not put Mélior personally in control of her marital destiny, but they are more compatible with the values that motivate her own desire. She wishes to reserve herself for Partonopeus precisely because of his noble aspect and refined demeanor. Indeed, her prognostication at an earlier moment in the narrative foreshadows Hernoul's:

> Ço ne lor seroit bon ne bel
> Qu'offrise à prendre un tousel;
> Mais ore quant venra li jors
> M'ert grans noblece et grans honors
> Que je vos puisse as miens mostrer,
> Qu'en vos n'aura que refuser;
> Car cevaliers eslis serés,
> Et sai très bien, jà n'i faurés,
> Et plus beaus que n'est riens el monde. (vv.1489–97)

> [*They would certainly find it ill advised should I propose to marry so young a bachelor. But later, when you are of age, it will be a great honor for me to be able to introduce you to my court. In you there will be nothing to disapprove of, for your will prove to be a knight of distinction (I have no doubts about that), and you will be more handsome than anything in the world.*]

In the comments that follow, the opposition of values evinced in Hernoul's speech manifests itself in more emphatic terms as Mélior describes Partonopeus's virtues a bit more explicitly:

> Qar toute beautés vos abonde,
> Et tant avés gentelise,
> Jà ne lairai ne vos eslise;
> Car vos estes del sanc Hector,
> Qui ainc n'ama argent ne or,
> Ne rien fors seul cevalerie. (vv. 1498–1503)[11]

> [*For you have abundant charms, and you have such exceptional refinement. Never will I allow them to choose any other but you. For you are a blood descendent of Hector, who never coveted silver or gold, or anything other than pure chivalry.*]

Therefore, although the critique of aristocratic marriage practice is rooted in evolving Church doctrine, we recall here the remark made by Colish, namely that the development of a sacrament relating to marriage could rarely be totally divorced either from the history of secular law and custom or from the dynamics of secular life and ideology. But to understand better the dimensions of this historic shift in values involving both religious and secular traditions, we need to examine more closely the relationship between the emergence of what is at least ostensibly a "Church" model of marriage and its competing "lay" model of marriage practice.

What we have begun to notice in our commentary on *Partonopeus de Blois* and *Flamenca* is that secular marital mores were not without their own conflicts and contradictions. In other words, the exploration of marriage themes need not rely upon a superimposed lyric tradition or theological canon, as such. Given the prominence of marriage as a real economic and political institution on the one hand, and as a sacred union on the other, secular experience and pagan legend provided an imaginative arena for implicit argumentation and speculation. We need to recognize that the very lyric *tropes* and theological formulas that influence the contemporary glossing of Classical epic and Celtic saga also figure as the implicit objects of speculation, definition, and debate.

In Gautier's *Eracle*, we do find recourse to a sage figure who provides a rather explicit wisdom motif, but this is exceptional. Even in the Old French *Tristan*, the introduction of the hermit figure Ogrin does not serve to introduce an explicit formulation of canonical marriage doctrine. In other words, we are not yet dealing with works like the thirteenth-century *Queste du saint graal* where religious hermits give explicit instruction in mystical doctrine to itinerant lay knights. In most of the works in our corpus, therefore, it is the larger secular conflicts surrounding the marriage dramas that provide the typological gloss. Consequently, our references to theological doctrine as a guide to the interpretation of Old French marriage fictions cannot stand independently of a more complex dialectic, to the extent that the very nature of the secular conflicts dramatized in extant literary genres or *exempla* provide an indispensable vehicle for the definition of emerging sacramental theories.

We have seen, for example, that chivalric values emphasizing individual honor and volition clash with alternate aristocratic values stressing paternal and corporate authority, and that the ideal of individual autonomy naturally converges with a Church doctrine emphasizing individual consent and affection. Furthermore, despite little or no explicit reference to theological perspectives, the Old French marriage dramas we have examined are consistently imbued with a deeper religious or divine dimension. We know that the courtship of Guilhem and Flamenca will take on pseudo-religious and ritualistic dimensions—the often satirical or fanciful nature of these correspondences would seem to assume the mature distillation of a standing tradition that associates amorous courtship with a divine mystery. Of course, the Classical *eros*, the Celtic *geis*, and the larger mythological tradition relating to the Indo-European sovereignty goddess and bride figure all serve to endow love and marriage with mystical dimensions. Obviously, a comparable association also obtains in the emerging twelfth-century definitions of the marital sacrament and its primary constituent, a willing spousal consent based on "love"—love as marital affection (*affectio maritalis* of Roman law), passive suffering (with its Classical models), and divine grace (effecting the operations of the Christian sacrament).

In *Partonopeus de Blois*, for example, we have seen an ingenious and subtle evocation of supernatural motifs. Later we will also see how the supernatural elements converge with a larger commentary on the religious dimensions of Partonopeus's marriage drama. We should also consider how the pathology of passive suffering inherent to the Ovidian model of love provides for parallels with the forms of religious experience dramatized in hagiographic narratives. First, however, I would like to turn to an analysis of an Old French epic work, *Girart de Roussillon*, so as to develop further our perspective on the conflicting secular ideologies that both construct and destabilize the private and social bonds relating to marriage within a larger feudal context.

The notion that early Old French epic inherently excluded erotic motifs finds its celebrated example in the Oxford *Chanson de Roland*. In other examples of the genre, representations of marriage tend to conform to Duby's lay model of feudal marriage practice, a practice dominated by political and dynastic prerogatives. But even a narrow focus on feudal institutions might be seen to provide for the immediate integration of erotic motifs to the extent that amorous devotion and fidelity harmonize with idealized forms of feudal loyalty between a lord and vassal—a correspondence which provides for a familiar metaphor in troubadour love poetry. This observation provides us with a point of departure for our present analysis of conventional epic motifs in *Girart de Roussillon*. This Old French epic poem from the second half of the twelfth-century reveals the extent to which exclusively secular ideological conflicts subtend the opposition of competing models of marriage practice. Despite the dominant role of epic motifs, amorous and religious themes also come to the fore as the poet attempts to account for a conventional conflict between a monarch and a rebellious vassal in terms of marital customs, affections and transgressions. Taken together, our remarks on *Partonopeus de Blois*, *Flamenca*, and *Girart de Roussillon* thus far will help us to understand the way in which marriage fictions integrated both secular and religious themes, as well as epic and hagiographic motifs, as a result of new theological and intellectual perspectives on marriage. We will see that the emergence of the Old French *roman* coincides with a profound shift in secular ideology and a concomitant reordering of heroic archetypes and ideological paradigms.

MARITAL CONSENT: *GIRART DE ROUSSILLON*

In his magisterial study *De l'histoire à la légende*, René Louis raises compelling questions about the origins and unity of the Oxford manuscript version of *Girart de Roussillon*. Although later critics have criticized Louis for being excessively exacting in his analysis of the poem's numerous narrative inconsistencies,[12] the fact remains that the important marital subplots that enhance the poem's epic action are almost entirely confined to the opening and concluding segments (commonly identified by the terms "prologue" and "epilogue") of a narrative that Louis judged to be the product of successive reworking by two or more anonymous hands.[13] The middle part of the narrative is built on motifs conventionally found in medieval epic, so it should perhaps come as no surprise that the amorous plots fade into the background. However, as Louis insistently argued, the middle portion of the narrative, which describes the diplomatic and military confrontations between the Frankish king and his rebellious vassal, contains virtually no remembrance of the elaborate marital intrigue that opened the story.

The most flagrant clue to the "patchwork" nature of the composition comes in the rather strained transitional *laisse* which marks the end of the poem's so-called prologue. In this passage the poet seems intent on anticipating the coming change in dramatic venue and eager to reconcile the amorous subplot with the epic action by announcing that a latent, amorous jealousy motivates the king's antagonism toward his most powerful vassal:

> Aisi duret tos tens l'amor des dous,
> Sanz nule malvaistat qui ainc i fous,
> fors bone voluntat e sanç rescous.
> E per hoc s'en fu Carles tan enviious,
> tot per autre auchaison ce li met jous
> An(z) fu au duc tan fers e tan irous,
> Per quan ferent batailles per plans erbou(u)s,
> Qu'en i at tant des mors, fei ke dei vos,
> Que li vif sunt restat tan tenebrous,
> Qu'ainc pois no fu parlaz moz amorous. (vv. 588–97)

> [*Thus continually the love between them both endured. Nor did either ever display anything other than good faith and secret loyalty. On account of this Charles became quite envious, but he quickly finds a false pretext for a challenge. Then he acted so proud and angry toward the duke that now they battle on the open field, and the fatalities, my word, are so great as to leave the survivors so downcast that no longer is there any talk of affection and reconciliation.*]

Louis's model of composite genesis gains further credibility when we recall that other passages in the "prologue" which fail to refer to the amorous intrigue already provide ample motivation for the ensuing conflict. Girart's considerable territorial power provides a significant threat in its own right to the king's sovereign authority, and this challenge to the king's power (a challenge exacerbated by the terms of the marital pact allowing Girart to gain almost total control over his most important *fief* by having it awarded to him in the form of an *allod*) should be enough to motivate the poem's epic confrontations.[14] Therefore, whatever position we might take regarding Louis's thesis, we still must ask ourselves how the anonymous poet or poets of *Girart de Rousillon* might have imagined the correlation between the poem's marital fictions and its central epic plot and themes.

It is interesting to note that even Louis, after persistently calling attention to the relatively marginal position of the amorous subplots, goes on to stress the critical importance of the love theme. For Louis, the love theme is second in importance only to the theme of social peace, but he argues that the poem praises what he calls *amour vertu* while condemning both sinful passion and matrimony:

> Le *Girart de Roussillon* du manuscrit d'Oxford se distingue et s'impose à l'attention par l'union de deux idées-forces: l'amour vertu, opposé aussi bien à la passion coupable qu'à l'union conjugale, et l'idéal chrétien de la paix tel que le résume la devise des fils de saint Benoît: *Pax*." (2: 416)

> [*The Oxford manuscript* Girart de Roussillon *stands out for its impressive linking of two seminal ideas: virtuous love (as opposed to guilty passion as well as the conjugal union) on the one hand, and the Christian ideal of peace on the other, as epitomized in the motto of the sons of St. Benedict: "Pax."*]

What exactly is *amour vertu* if it is neither a purely spiritual love, nor conjugal love, nor erotic love? When Louis speaks of a love opposed to "the conjugal union" does he mean a love at odds with marital vows, as troubadour love is often characterized for example? This is what Louis implies in his reference to the poem's "southern" origins and its apparent *troubadour* influences.[15]

Is there not an intuitive contradiction in Louis's evocation of anti-matrimonial bias in a twelfth-century poem that also envisions a Christian ideal of social peace and unity? How do the amorous intrigues and marital plots contribute to the poem's fundamental interest in the problem of feudal warfare and to its creative vision of social peace? In short, how do the amorous themes mesh with the central concerns of an aristocratic epic?

We have pointed out, for example, that in Duby's schema of medieval marriage practices the marriage ideology based on affection and individual consent appropriate to what he calls the "ecclesiastical model" is essentially antithetical to that of his second principal model, the "lay model" which governs secular customs (*Two Models* 12–17). With regard to this ideological opposition, Louis's comparison of the Brussels manuscript version of *Girart de Roussillon* with the Oxford manuscript version is illuminating (3: 73–75; 79–81). In the former version, there is no trace of the Fouque / Aupais subplot, and amorous sentiments are limited to what Louis describes as a kind of "fraternal affection."[16] As evidence of the "lay" ethos governing the Brussels manuscript version, we can point to Louis's observation concerning the double marriage at the beginning of the poem: Charles and Girart arrange a final settlement independently of their prospective fiancées, and the terms of the marital swap are reduced to those of a mundane contract of material exchange negotiated without the consent of either fiancée. In the Oxford manuscript version, on the other hand, Girart is repulsed by the idea of reducing the conflict to economic calculations, and he initially refuses any material compensation. Rather, he seeks (as we shall see) to reaffirm his moral honor and personal prestige. He also solicits Elissent's approval before making any pledge to Charles. Finally, the

Oxford manuscript version also lends an amorous significance to the exchange-of-rings motif (in the Brussels manuscript version, by contrast, the ring is introduced only as a device for authenticating the veracity of a message sent by Elissent to Girart).

It is clear, therefore, that the depiction of marriage in the Oxford manuscript version departs dramatically from the conventional representations of marriage in feudal epic and from their corresponding features in Duby's "lay" model of twelfth-century marriage ideology—a model in which paternal authority and political interest exclude any notion of mutual affection and consent.[17] Perhaps it is precisely this divergence which encouraged Louis to define love in contrast to marriage and to discern an anti-matrimonial sentiment in the poem. However, both of Girart's marriages feature ties of real mutual love in conjunction with binding matrimonial vows that in no way exclude an enduring amorous sentiment.[18]

We have seen that the depiction of an aristocratic marriage in *Flamenca* involved an ironic viewpoint informed by the tenets of the consensual model of marriage. Likewise, the vision of love and marriage in *Girart de Roussillon,* which diverges from the "lay model" of marriage defined by Duby, shares many features highlighted by Duby's "ecclesiastical model" of twelfth-century marriage practice. Here again, however, we must contend with the limits of Duby's dichotomous scheme. While the tenets of the new Church doctrine could be construed as conferring a new legitimacy on clandestine marriages and on marriage vows made independently of paternal authority, it is also true that canonists tended toward a certain moderation in all of these matters. Clearly, the Church could be seen as a twin pillar of Western power and not solely as an ideological adversary of the secular nobility. The papacy had its own corporate interests to protect and an inherent suspicion of any radical concept of individual autonomy (Duby *Two Models* 21–22). In the end, therefore, it becomes difficult to maintain a strict division between the "lay" and "ecclesiastical" models of twelfth–century marriage practice defined by Duby.[19]

We find this kind of ambivalence with regard to ecclesiastic and secular attitudes in *Girart de Roussillon* culminating in a broader critique of feudal ideology informed by emerging marriage doctrine and targeting papal and aristocratic practices alike. While a profound religious sentiment pervades the poem, especially with regard to Berthe's spiritual devotions and the evocation of Mary Magdalene, some Christian motifs and conventions prove to be flawed and ineffectual. In the realm of war, it is noteworthy that the miraculous burning of the standards—a divine miracle provoked by the bloody battle at Vaubeton between Girart and Charles—fails to secure a lasting peace. Likewise, the Pope's role in the betrothal of the Emperor's daughters will ultimately prove ineffective in assuring a stable domestic peace and a secure imperial alliance.

The Pope's nearly frustrated attempt at reconciling Girart to the marital swap capriciously imposed by the king provides us with a revealing insight into the close relation between marital intrigue and the poem's broader critique of feudal institutions. The Pope wrongly assumes that a generous material benefit will suffice to placate Girart for the damage done to his honor and for the loss of his intended fiancée (which, by the way, is precisely the kind of exchange agreed upon by Girart and the king in the Brussels version where we saw a closer conformity with Duby's "lay model"):

> Per la fei que dei Deu omnipotent,
> Mais preis assaz Bertan que Alisent.
> Cons, vai, pren la muller e tot l'argent,
> Les chevaus e les pailes e l'ornement.
> S'en vols onor ne terre ne creisement,
> E Carles en fera tot ton talent,
> Aisi com el m'a dit, s'el ne m'en ment. (vv. 431–7)

> [*By the faith I owe almighty God, Berthe surpasses Elissent in worth. Go count, take the woman and all the money, the horses, the fine clothes and furnishings. If you wish to procure further honors, land, or income, Charles will grant your very wish, just as he swore to me, if he is not lying to me.*]

The Pope's exhortations reveal that he remains blind to the nature of the sentiments that actually motivate Girart and fails to appreciate the *moral* and *affective* nature of the injury suffered. This gap in perspective is made immediately apparent by the reaction of Girart's personal advisors:

> Ce lauent a Girart cil sui parent:
> Toz ert honit li cons s'aver en prent;
> Mais pur son fiu li solve tan cuitement
> Si qui li cons ne tiegne de lui nient,
> Ne mais ses om ne sie a son vivent. (vv. 438–42)

> [*Girart's kinsmen advise him of their opinion: "If the count accepts any material gain, he will be disgraced. But for his fief he should be fully absolved, such that the count owes him nothing and will no longer be his vassal as long as he lives."*]

How do we explain this subtle but important divergence in attitudes? The Pope fails to realize that from Girart's perspective Charles has violated another kind of "honor," namely the "honor" of his word—that is the (good) faith ("foi") that truly ratifies the rites and promises that bond vassal and lord. The ideological gap separating Girart (along with his advisors) from the Pope is reflected in the shifting semantic value of the word *onor* and calls attention to the semantic ambiguity of a word whose meaning is subject to a historical slippage between its purely *material* meaning and an emerging *subjective* or *moral* meaning.

The Pope's materialistic perspective and lack of diplomatic savvy underline the critical role played by subjective "honor" and "good faith" in relation to the feudal contract. In twelfth-century feudal discourse, an honored promise is a sign of good *faith* (*foi*), and the violation of this faith is formally denoted by the verb *mentir* (Matoré 142). Thus, the "deal" formulated by the Pope and forged for the purposes of reconciling baron and king—a reconciliation upon which the fate of the Christian alliance eventually depends—calls attention to the fundamental inefficacy of its own terms. The material offer of "onor" does not address the offense suffered, and the Pope's last qualification ("selon ce qu'il m'a dit, s'il ne m'a pas menti") underscores the fragility of the entire proposal. In fact, mendacity is one of the king's salient vices. It is ironic that the Pope, whose office assumes consummate statesmanship, would fail to see that the absence of good faith on the part of France's sovereign condemns the papal settlements to utter futility. The poet's critique is made perfectly clear by the ensuing military confrontations—confrontations which the papal diplomacy failed to circumvent.[20]

Two facts have now been made evident: first, it is difficult to assess the significance of the marriage plots if we rely entirely upon the opposition between the "ecclesiastical" and "lay" models inherent to Duby's theory of medieval marriage; second, the depiction of marital affection and consent in the Oxford version of *Girart de Roussillon*, while it may reflect a reworking of an extant epic tale and a "gloss" influenced by theological marriage doctrine, finds its own inherent significance in direct relation to formative ideological tensions within secular institutions. In order to appreciate better the relation between the poem's parallel feudal and marital critiques, I suggest an alternate model of marriage practice. In his work on Indo-European marriages, Georges Dumézil defines three ideological categories with respect to his fundamental tripartite division of Indo-European society. The essential traits of each model of marriage practice can be summarized as follows:

> *First Function*: Marriages between members of the ruling or priestly class based on formal protocol and carefully observed rituals. (In Duby's paradigm, by contrast, these traits might be shared by both of the two opposing models).

> *Second Function*: Marriages established according to the autonomous will and desire of the private individual or as a result of the autonomous desire of two freely and mutually consenting individuals (elopement, feigned abduction). Emphasizing freedom of action and freedom from external social constraint, this category is associated with Dumézil's second function or military caste and (because individual volition can manifest itself unilaterally as rape or abduction) proves to be the most ethically problematic

model. (We notice that some of these same concepts are, paradox-ically, reflected in Duby's "ecclesiastical model" because of the importance of individual consent in medieval canon law.)

Third Function: Marriages based on economic contracts and equal socio-economic status and which conform to the rules of econom-ic exchange. This third model of marriage originates with agrarian and mercantile castes. (We should note that property interests form an important feature of Duby's "lay model".)

While rudimentary and static as a paradigm, we cannot help noting how intuitively different the Dumézilian schema is from Duby's binary model of twelfth-century marriage practices. Dumézil's model allows us to identify an inherent logical correspondence between Girart's rank in a secular, feu-dal hierarchy and his particular attitudes toward marriage. In both Dumézil's second function ideology of marriage and twelfth-century eccle-siastical doctrine on marriage freedom of choice and individual desire and affection are central values with regard to courtship and marriage. One can therefore identify a common ideological term providing a link between a) the secular and second function origins (warrior class) of a *consensual* model of marriage and b) the emerging theological doctrine of consent adumbrated at a secondary level of representation. Without this conver-gence of feudal and religious ideology the narrative perspective would require, by necessity, recourse to an explicit statement of religious teaching apparently with no real anchor in the temporal conflicts surrounding what is, however it may be endowed by spiritual teaching, a fundamentally sec-ular institution. One might also speculate that the emerging theory of mar-riage as a sacrament gains a further sanction by virtue of a narrative action which highlights the *universal* laws of consent and affection. Finally, the tensions and conflicts inherent in (pagan) aristocractic custom provide an argument in favor of a transcedent Christian legislation and ethos—one which gives explicit and perfected expression to the universal laws already revealed in secular affairs.

Dumézil's tripartite model also has the advantage of defining marriage practices according to broad categories of Indo-European social ideology that translate into other domains of social practice. From this perspective, the implicit justification of a sacramental theory of marriage and the con-trast with flawed aspects of customary marriage practice converge with a more inclusive critique of extant feudal society. We can, for example, iden-tify important parallels between Dumézil's model of Indo-European mar-riage and Le Goff's functional analysis of twelfth-century feudal vassalage. Le Goff (352–73) identifies three fundamental components to the bond between lord and vassal:

The pledge of "homage" (*hommage*) by the vassal symbolically and effectively affirms (in rite and function) the superior rank and authority of the lord.

The mutual pledge of "faith" (*foi*) ratifies and protects the integrity of the feudal relationship and has as its peculiar characteristic a purely moral aspect based on the presumed nobility of each party. Le Goff emphasizes that this equal nobility and its concomitant rituals affirm (somewhat paradoxically) the moral *equality* of both lord and vassal. Just as the emphasis on personal autonomy makes Dumezil's second-function marriage customs ethically ambivalent and problematic, so too does this inherent subjective equality threaten to subvert the hierarchical divisions that define feudal bonds of vassalage.

The final component is the *don* or "gift"—the benefits and rights to land tenure received by the vassal in exchange for the pledge of homage. This component underscores the *economic* obligations to be fulfilled by both parties.

Le Goff's model serves as a reminder that over the course of centuries various aspects from all three Dumézilian categories are likely to reveal themselves in the synthetic rites and attitudes of a single class. This does not mean, however, that these distinctions have faded from sight. In fact, Le Goff emphasizes the contrary:

> Il faut enfin bien remarquer que si l'hommage, la foi et l'investiture du fief constituent un système unique et complet, les significations des rites symboliques successifs ne se détruisent pas mais s'ajoutent. (371)

> [*Finally, it is well worth noting that if the exchange of homage and faith, and the investiture of the fief constitute a unique system unto itself, the meanings of the successive symbolic rites do not efface one another but are joined together.*]

Even more importantly, however, Le Goff insists on the indispensable *integration* of all three components in accounting for the unique features and proper functioning of the "système féodo-vassalique:"

> Les rites de vassalité tels qu'on les observe dans la société médiévale occidentale constituent en effet un *système symbolique global* et ce système est *original* . . . aucun des deux modèles économiques principaux de réciprocité qu'offrent les sociétés préindustrielles [le potlatch, le droit romain] ne semble pouvoir s'appliquer au système féodo-vassalique. (372)

> [*The rites of vassalage as we observe them in western medieval society effectively constitute a* global symbolic system, an *original* system . . . . *neither of the two principal economic models of reciprocity in pre-indus-*

trial societies (potlatch and Roman law) would seem to prove applicable
to the system of feudal vassalage.]

In *Girart de Roussillon* the poet comments critically on the contribution of each part to the harmonious functioning of the social institutions and on the inherent tensions and paradoxes that plague the effort to integrate them. Le Goff's economic and ethnographic observations reveal not only the rather predictable correlation between aristocratic feudal concepts and an aristocratic marriage practice based on principles of paternal authority, economic exchange, and contractual obligation, they also allow us to see that the "competing" marriage model (based on spousal affection, subjective honor, and consent) need not *necessarily* be derived from a distinct regime (such as lyrical poetry or sacramental theology) because it has its own basis in essential features of secular class divisions and ideologies.

Together, the models of Dumézil and Le Goff allow us to construct a deeper ideological vision of medieval marriage as an institution which is shaped by corporate and nationalistic interests, which crosses the line between secular and religious domains, and which makes marriage into a symbol of entrenched prerogatives and as a site for challenges to established hierarchies. This provides ripe terrain for both a dramatic and critical elaboration of otherwise purely theological formulations on the role of marriage in Christian society.

It is interesting to note, for example, that in the models of both Dumézil and Le Goff, the *affective* and *subjective* nature of volition (second category in both systems) plays a key but problematic role in relation to the other forms or components. The distinctions provided by Le Goff's tripartite model allow us to anticipate the failures that plague the Pope's diplomatic maneuvers while also explaining the logic behind the convictions that animate Girart. That Girart is not seeking to profit in a *material* sense by the appropriation of his land in the form of an *allod* is clear from his anger over the compromise he is forced to adopt ("Ab aiqueste paraule Gerrarz s'enprent" v. 443), and he remains resistant even to the exhortations of his own kin and counselors (vv. 450–1). However, the agreement provides Girart with a means for addressing his wounded pride as well. For, as his counselors imply, the king's broken pledge results in releasing Girart from the ties of vassalage that implicitly affirm his subservient position *vis-à-vis* the king (vv. 438–42, cited above, p. 9). In yet another show of concern for his honor, Girart insists upon first "testing" Elissent's good faith, lest she too profit cynically (by her marriage to the French king) at the expense of his own reputation. Once he obtains proof of her sincerity and integrity, he responds with a demonstration of reciprocal respect and service as demonstrated in this exchange with Elissent:

Cui volet melz, donzele, mei o cest rei?
— Se Deus m'ajut, charz saigne, eu am plus tei.
— Se m'agu[i]ssaz orguel dit ne desrei,

Ja mais ne vos tangest dojoste sei.
Er le [le roi] prendez, donzele, eu t'ou autrei;
Eu prendrai ta seror per amor tei. (vv. 463–8)

[*"Whom do you prefer more, lady, me or this king?"* —*"God help me and
cross my heart, I love you more."* —*"If you had said anything presump-
tuous or impertinent, I would never more stand by your side. Now, marry
the king, lady; I give you leave to do so. I will marry your sister out of my
love for you."*]

Thus, the amorous elements of the marital subplot issue naturally from the
poem's preoccupation with the role of affective bonds and subjective honor
in feudal alliances and contracts, and Mary Hackett has argued that the
language of love and the use of the word *amor* in the poem find their closest parallels in the customary and sober terminology of the feudal contract
between vassal and lord (729–30). Indeed the notion of subjective honor
and faith (i. e. as "good faith" and honest intention) is a common element
not only to feudal contracts but also to emerging twelfth- and thirteenth-
century ethical thought.[21]

By referring to Dumézil's Indo-European model of marriage practices,
we likewise find a key to the fictional models underlying the Oxford version's parallel elaboration of martial and marital conflict and to its elaboration of a rather peculiar double-marriage motif. Within the broad outlines of Indo-European sovereignty myths, we find a confluence of narrative elements that also appear in *Girart de Roussillon*. Selecting only the
most prominent of a large number of parallels, we can point out that 1) the
intimate connection between a marital and sovereignty rivalry, 2) the hero
who travels to foreign lands on a bride quest and who is recommended by
his past exploits, 3) the combined motifs of destitution, exile and restoration in combination with a powerful bride figure, and 4) the double bride
motif—all of these appear in *Girart de Roussillon* and have parallels in
Indo-European sovereignty myths.[22] The parallels between the exile
episodes in *Girart de Roussillon* and in the *Tristan* legend also suggest the
influence of Celtic models (widely recognized as valuable repositories of
archaic pre-Christian traditions). Finally, the marvelous aspects of Aupais's
role only seem to underscore further the poet's apparently conscious
exploitation of a pagan subtext.

While the poem's marital dramas reflect the influence of specific twelfth-
century doctrines on marriage, it is clear that the outline itself of the epic
tale of *Girart de Roussillon* also provided a vehicle for understanding and
defining the very logic of such doctrines. Just as the initial marital arrangement (invalidated by the king's excessively sensual capriciousness) enacts a
drama of medieval sovereignty by leading to a violent political rebellion
and by threatening the larger political alliance between the Frankish and
Byzantine realms, so too does Girart's experience of personal suffering
reveal in its Tristan-like modulations the tragic drama of the individual

whose marital destiny is thwarted by secular customs and impositions.[23] Finally, the Fouque / Aupais subplot stages an alternate ideal. This idyllic fairy tale of enchanted love between a warrior baron (Fouque) and his jail keeper (Aupais) evolves into a valid marital union that defies and overcomes a deeply entrenched rivalry between antagonistic clans. Accordingly, it is Fouque's star which is rising at the end of the narrative. The sacramental nature of marriage reveals its divine image in Girart's hagiographic itinerary but seems to fulfill its true end as an institution of secular peace, renewal and reconciliation in the marriage between Fouque and Aupais. A progressive theological gloss superimposed on epic and pagan models finds in the drama of love and marriage a fictional topos for envisioning a new and affective form of religious and feudal "faith" as the basis for a lasting social peace.

In examining marriage motifs in *Girart de Roussillon*, it is very clear that the representation of marriage and the interest in marital and adulterous conflicts in Old French narrative involves much more than the influence of contemporary Provençal lyric tradition. It is also clear that the notion of consensual marriage plays an important role and that a dichotomy exists which essentially conforms to the opposition proposed by Duby, between lay and ecclesiastical models of marriage. On the other hand, the poem clearly reveals that the ecclesiastical model responds to and converges with ideological conflicts already inherent in the aristocratic (lay) norms, norms whose coercive features must almost inevitably engender profound social conflict. We can argue that the concept of *amor* provides a nexus which links basic features of feudal society—namely the *affective* bonds of love, loyalty, and faith critical to the institution of vassalage. But here the feudal elements also resonate with contemporary doctrine on marriage, to the extent that these affective bonds are also those that define the Christian subject (in relation to the operations of divine grace) and that (ideally speaking) help to create a feudal state in harmony with a higher providential order. Therefore, the epic motifs and the imposed Christian gloss readily complement one another such that the didacticism of the Old French *roman* reflects a subtle transformation of secular values that privileges, somewhat paradoxically, neither the titular ruler (first category in Dumézil's tripartite model of Indo-European social structure and ideology), nor the warrior caste (Dumézil's second category), but rather the amorous hero whose relationship to society and to history is defined by affective bonds and amorous conquest.

TRANSLATIO IMPERII

Since emerging doctrine makes marriage the defining institution that serves to order a unique secular estate, the literary representation of marriage fortunes will readily express itself in terms of what are ostensibly secular (political, historical, or feudal) concerns and secular values. In many of the works in our corpus, marriage plots are intimately linked to the narrative topos of *translatio imperii*. Even in the epic context of *Girart de Roussillon*, the hero (following his period of defeat and exile) secures the restoration of his feudal *allod* by virtue of his affective relationship to Elissent, whose manipulations lead the king to give back to Girart his former domain at Roussillon. In our analysis of two *romans d'antiquité*, we saw that marriage dramas, inflected by a ubiquitous Indo-European sovereignty and sovereignty-bride mythology, also played a defining role in terms of imperial conquest and national history. We need only compare these "epic" narratives to the Oxford *Chanson de Roland* and its accounting of Charlemagne's imperial conquests to see the significant shift in secular values played out in relation to marriage fictions in Old French literature.

Even in a work such as *L'Escoufle*, where the lovers' itinerary is clearly inflected by hagiographic models (to be discussed below), the final triumph is ostensibly political and feudal and reflects, once again, the nearly ubiquitous relation between marital plots and *translatio imperii* in the works in our corpus. Guillaume's ostensible marital "misfortunes" seem first to deny him succession to the Roman empire, leading into poverty and an exile that ends back in his native French domain. In France, Guillaume gains the affection of the aristocracy in Provence, and then assumes his rightful place as a count in Normandy subsequent to his reunion with Aélis. He immediately earns the love and devotion of his subjects who fondly remember the generous rule of Count Richard, Guillaume's now deceased father. But praise for Guillaume and Aélis is so extraordinary that ultimately news of their triumph makes its way back to Rome, and Guillaume is invited to succeed the late Roman emperor (vv. 8512ff.).

Whereas the epic works we have examined (*Roman d'Eneas, Roman de Troie, Girart de Roussillon*) make imperial conquest a central motif, a work such as *L'Escoufle* tends to be preoccupied with love itself, as a formative feature of private and public social harmony. At the end of the tale, Guillaume's relation with the imperial court is no longer depicted in terms of the rival interests that conspired to separate him from Aélis, and his succession to the imperial throne is depicted not as a political, national, or military conquest but as a universal jubilation involving amorous flirtation and courtly generosity (vv. 8768ff.).

We find a similar transformation of epic motifs and aristocratic norms elaborated in *Partonopeus de Blois* as the ostensible dichotomy between the "real" or normative secular court (Frankish court) and the "fairy" court (Mélior's court) is ingeniously deconstructed. Partonopeus's first encounter with Mélior strikes the reader as an encounter with metaphor itself: Mélior and her infinite kingdom seem to be the idealized fantasies of a youthful libido. When the encounter evolves into a vulgar rape, however, the fantasy is quickly deflated and is seen to be nothing more than the obverse of a normative aristocratic reality based on patriarchal forms of coercion and economic conquest. Yet, by virtue of the amorous suffering which dominates the lovers' subsequent experience and ultimately secures their marital desires, the metaphoric value is recuperated. While Partonopeus's rivalry with other suitors during the course of the tournament underlines the economic and territorial stakes at hand, it remains clear that for both Partonopeus and Mélior (and from the reader's perspective) the political import is diminished in relation to the psychological suffering of the characters, which has been elaborated with much sentimental drama. The marital itinerary does indeed determine a political fortune, but for all intents and purposes the political and economic conquest regains its value as a psychological and imaginative metaphor—a metaphor that conveys a deeper and visceral affective relief and triumph, the consummation of an idealized and long-deferred amorous destiny. Here the notion of imperial triumph is ultimately much less literal than it is, for example, in Enéas's conquest of Latium or in the transfer of empire from Troy to Latium. Love, as the very vehicle of *translatio imperii*, is gradually displacing martial and imperial ideology during a period when the figures of Paris and Tristan become primary archetypes.

TRANSLATIO IMPERII: CLIGÈS

The extent to which twelfth-century authors consciously linked marital fortune with the topos of *translatio imperii* is made most explicit in Chrétien de Troyes's *Cligès*. Not only does the eventual resolution of the political conflict between Cligès and his cousin Alis (wrongful usurper of the Greek imperial throne) parallel the deferred consummation of Cligès's marriage, but the author frames the entire tale with explicit references to the topos of *translatio imperii*. The well-known prologue to the tale not only makes clear that the author is defining precisely what *translatio imperii* denotes— namely a kind movable manifest destiny—but it also defines the ideological orientation of the work by distancing it from pseudo-historical and propagandistic pretensions. This is to say that the notion of a national manifest destiny is defined here in terms of cultural achievement and renown (*chevalerie* and *clergie*), not in terms of a territorial conquest as such:

> Ce nos ont nostre livre apris
> Qu'an Grece ot de chevalerie
> Le premier los et de clergie.
> Puis vint chevalerie a Rome
> Et de la clergie la some,
> Qui or est an France venue. (vv. 30–35)

> [*Our books have taught us that chivalry and learning first had the greatest fame in Greece. Then Rome became the center for them both which now have come to France.*]

By comparison to works such as the *Roman d'Eneas* or *Partonopeus de Blois* where the larger frame of reference involves a fictional genealogical affiliation between a contemporary court and a former imperial nation, Chrétien gives *Cligès* a novel orientation with regard to the Classical past. Indeed, Chrétien actually emphasizes the distance between past and present by questioning the contemporary relevance of a past culture and past doctrines:

> Car des Grezois ne des Romains
> Ne dit an mes ne plus ne mains:
> D'ax est la parole remese
> Et estainte la vive brese. (vv. 41–4).

> [*For nothing at all is said anymore with regard to the Greeks and the Romans. The living ember of their fame is extinguished, and one cease to hear of them.*]

For as the very name of the authorial persona suggests, the triumph of the contemporary age is necessarily a Christian one; it is a triumph of a living Christian spirituality and doctrine. Hence, Chrétien's work does not emphasize the providential import of marriage in terms of literal marital conquest, territorial domain, and genealogical descent, but rather in terms of marital practice itself and its role in relation to social cohesion. Less important than who wins the bride, is how society (by virtue of its mores) defines who marries whom. The Arthurian setting is fitting in this regard. It provides an imaginative space that is not Greek or Roman and which emblematizes (in the persona and exploits of Arthur) feudal cohesion. Obviously, however, it is significant that this social idyll is elaborated in terms of marriage and courtship dramas, suggesting once again the influence of a theory of marriage in which marriage plays a pivotal role in relation to secular culture.

Unlike Benoit's tragic and epic view of conjugal strife, for example, Chrétien's work is a consummate *comédie de moeurs*, but no less explicit as a critique of amorous and marital mores. At the end of the text, for example, the *translatio imperii* conceit that frames the beginning of the tale is ingeniously recuperated by virtue of an almost burlesque historical footnote that serves as a kind of closing *moralité* to the elaborate intrigues and deceptions that precede Cligès's eventual triumph over the usurper Alis:

> Et chascun jor lor amors crut,
> Onques cil de li ne mescrut
> Ne querela de nule chose,
> N'onques ne fu tenue anclose
> Si com ont puis esté tenues
> Celes qu'aprés li sont venues ;
> Einz puis n'i ot empereor
> N'eüst de sa fame peor
> Qu'ele nel deüst decevoir,

Se il oï ramantevoir
Comant Fenice Alis deçut
Primes par la poison qu'il but
Et puis par l'autre traïson.
Por ce einsi com an prison
Est gardee an Costantinoble,
Ja n'iert tant haute ne tant noble
L'empererriz, quex qu'ele soit :
L'empereres point ne si croit
Tant con de celi li remanbre. (vv. 6743–61)

[*And each day their love increased; never did the one mistrust the other. Nor did they quarrel about anything; never was she sequestered—such as those were who came after her. For from then on there was never an emperor who didn't fear that his wife would not deceive him once he heard the story—how Fénice had deceived Alis, first by the potion he drank, and then by that other deception. Thus, on this account, as if in prison, she is kept under watch in Constantinople, however powerful, however noble the empress. Whoever she may be, the emperor does not trust her as soon as he remembers the tale of this earlier one.*]

Hence, marital mores define cultural heritage, just as that heritage (i.e., *clergie et chevalerie*) governs the transfer of dominion (*translatio imperii*). In the case of contemporary Greece, however, the heritage is (by Western standards) not only exotic but farcical—as a reflection and extension of the kind of romantic intrigues and excesses that prevailed in the courtship of Cligès and Fénice:

Toz jorz la fet garder en chanbre
Plus por peor que por le hasle,
Ne ja avoec li n'avra masle
Qui ne soit chastrez en anfance:
De ce n'est crienme ne dotance
Qu'Amors les lit an son lïen.
Ci fenist l'uevre Crestïen. (vv. 6762–8)

[*Every day he keeps her sequestered in her chamber, more out of fear than because of the sun. Nor will any man ever be in her company who has not been castrated as a youth, so that there can be no fear or suspicion that Love bind them in his net. Here ends Chrétien's tale.*]

Chrétien's text underlines the fluid nature of emerging lyric and theological views on love and marriage. The ironies and the comic peripeteia have an alienating effect, distancing listeners from pathetic responses. Instead, the prologue invites the reader to envision the relative value of private actions within the context not of religious moral prescription as such but as a mirror of a larger, potential cultural achievement.[24] In this sense, *translatio imperii* in *Cligès* and other twelfth-century marriage dramas sets in relief a vision of secular culture as a rational order and construct in its own right (precisely the order served by the marriage sacrament) and as the index for assigning value to private (lay) action and "morality" rather than the other way around, rather than making conformity to patristic moral prescriptions the explicit measure of secular achievement.

MARRIAGE DOCTRINE AND THE SECULAR ORDER

In fact, Chrétien's comic invocation of the *translatio* motif at the end simply refers us back to the tale itself, to its didactic import and to the theme of marital consent in particular. Indeed, the ending forces us to contemplate the characteristics that differentiate the courtship of Cligès and Fénice from that of Alexander and Soredamors. In presiding over the latter marriage, Guinevere delivers what is clearly Chrétien's most explicit and succinct didactic statement in a work that is frequently ironical, farcical or burlesque:

> Or vos lo que ja ne querez
> Force ne volanté d'amor.
> Par marïage et par enor
> Vos antraconpaigniez ansanble. (vv. 2286–9)

> [*Hence I advise that you never seek willful or coercive love. Better you should live together in marriage and in honor. Only this way, as it seems to me, will your love be able to endure.*]

Although the marriages of both Alexander and Cligès to Soredamors and Fénice respectively involve an initial infatuation and an enduring affection, it is clear that amorous desire is to find its resolution in an open public rite that confers proper entitlements and benefits. But the specific terms of the negative admonition are unusual: *ne querez /* [*ni*] *Force ne volanté d'amor*. In their 1993 edition of *Cligès*, Gregory and Luttrell point out that critics have previously proposed the following interpretation: "(Je vous conseille de ne jamais chercher) de résistance à l'amour, ni de complaisance pour l'amour." In response, they go on to point out the obvious fact that "On a affaire à une condamnation d'une forme d'amour opposée à celle caractérisée *Par marïage et par enor*." Accordingly, they suggest: "Le conseil de la reine consiste à ne pas rechercher du tout une liaison amoureuse, en cédant à la force irrésistible de l'amour ou en choisissant de leur propre gré d'être amants."[25] This rendering implies a too wholesale condemnation of love, however, given that both marriages are based on an initial amorous infatuation. Of course there may also be a deceptive irony in the apparently prescriptive style of the queen's admonishment, an irony which defeats its didactic formality. For, Guenivere's authoritative tone may reflect not so much impartial reason as it does either a profound conviction or a shameless hypocrisy—with regard to the memory of that adulterous transgression accounting for her own fame. It is difficult to be sure of Chrétien's intent, but clearly such a reading increases the distance between the didactic statement and the authorial voice or persona.

In any event, the queen's prescription seems to involve subtle semantic distinctions. However, our earlier remarks on twelfth-century definitions of consent may provide some insight. We recall that the term *volition*, synonymous with the term *consent*, was eventually displaced by the latter term as Christian thinkers sought, so it appears, to exclude from the notion of marital consent any radical expression of autonomous individual volition (see Chapter Three). We have seen, however, that Old French representations of marriage also reflect tensions deeply embedded in secular ideology as well, and that Christian doctrine converges with an undercurrent of lay attitudes based precisely on an inherent aristocratic regard for individual freedom—a conflict we saw elaborated in the epic poem *Girart de Roussillon*. However, Girart's relentless and ultimately destructive pursuit of political and amorous autonomy highlights the problematic ambiguity inherent to Dumézil's second category marriage model, in which the high regard for individual autonomy may as readily converge with a canonical notion of consent as it may manifest itself in terms of violent abduction or individual rebellion. In the case of *Partonopeus de Blois*, for example, the hero's "romantic" affair with Mélior provides a model of courtship and marriage which is essentially antithetical to the aristocratic model. Yet the initial encounter is marked by the alternate aspect of second category prerogatives—namely, a unilateral act of willful aggression. In this case, the act takes the form of Partonopeus's rape of Mélior. It is within the parameters of this problematic ambivalence between the consenting and the lawless subject that one must attempt to render the meaning of Chrétien's prescription as well as the work's intertextual references and responses to the Old French *Tristan*, to *fin'amor*, and to other twelfth-century perspectives on love and marriage. Clearly, Chrétien was aware of the dangers inherent in pagan models. Whatever his exact positive prescriptions, he seems bent on admonishing the reader who too readily sees the cultural achievement of a Christian France in a direct line with temporal and secular views as such—lest the reader forget: "Des Grezois ne des Romains / Ne dit an mes ne plus ne mains: / D'ax est la parole remese / Et estainte la vive brese."

Ultimately, therefore, the Old French *roman* seems to define a new social space that is distinctly secular—involving explicit interests in love, vassalage, sovereignty, and marriage—but, by virtue of the sacramental role of marriage, not entirely divorced from Christian providence and Christian moral doctrine. Nonetheless, the very interests which define temporal value and the itinerary of the secular individual—like the marital imperative itself—distinguish the temporal life quite fundamentally from that of the monk or cleric. We saw, for example, that the Oxford manuscript version of *Girart de Roussillon* testifies to the emergence of a new spirituality based on more affective and subjective categories of love, ethics, and religious faith. By involving both secular and ecclesiastic structures (i.e., monarchy and papacy) on a single plane in relation to the struggles of its

central protagonist, the poem also reflects the need to redefine the forces
and norms that determine the fortunes of secular subjects. Girart's itinerary, rather fatally circumscribed by corrupted feudal structures and norms,
reflects the limits of the "old" dichotomy (a dichotomy between secular
perdition and absolute religious seclusion and redemption), while the
Fouque / Aupais marital intrigue leads instead to a symbolic erasure of
entrenched feudal hatreds and vendettas. In contrast to the clear patristic
oppositions between "sin" and "virtue" as between "religious" and "secular, "divine" and "worldly" therefore, the new order also involves certain
paradoxes. The new dispensation provides for a view of secular life that
goes beyond the mere possibilities of private moral conduct (with pastoral
approval or disapproval) and which involves the *secular* subject with a
Church sacrament unique to the temporal "estate." This ideological development can also not be isolated from western literary history, from the
influences and novel adaptations of amorous lyricism, of the Celtic marvelous, and an enduring Indo-European sovereignty and sovereignty bride
mythology. At the center of this erotic typology (going back to Ovid and
Virgil) one finds a depiction of love that is both passionate and divine and
that is resurrected most dramatically in the seminal motifs of the Old
French *Tristan et Iseut*.[26]

MARRIAGE DOCTRINE AND THE SECULAR ORDER: *TRISTAN ET ISEUT*

Indeed, the Old French *Tristan* narratives provide a centerpiece for two
centuries of French vernacular literature dominated by questions involving
love, marriage, and adultery. Sarah Kay has noted, however, that the Old
French *Tristan* poems are equally "elusive" as to genre and ethos as they
are influential with regard to theme and action. I believe our insights into
Old French marriage fictions can help resolve some questions of genre and
ethical perspective presented by the works of Béroul and Thomas and that
this analysis will in turn prove helpful in understanding formal features and
ethical questions in subsequent marriage fictions.

Kay points out that while two ethically problematic motifs found in the
Tristan story, namely adultery and the duplicitous oath, find their way into
other Old French secular narratives, the amorous quest more clearly
resolves itself with regard to action and ethos in later works than it does in
the *Tristan* narratives. Kay argues that in Béroul's version, for example, the
lovers seem to relish practicing deception and show little regard for social
laws. In both Béroul and Thomas the lovers also grow increasingly alienated from the society around them (Kay 190–193). Le Gentil has contend-
ed that Béroul's narrative intentionally obscures any divinely authoritative
perspective and judgment on the lovers' actions ("Tristan" 115–116).
Subrenat and Hunt have both argued that contemporary ideas allow us to
resolve such apparent ambiguities. Indeed, in an article published in 1976,

Subrenat makes a clear case for reading Béroul's version in the light of twelfth-century doctrines on consensual marriage (225–35). Of course Subrenat's observations, like our own discussion of emerging marriage doctrine, help us to explain Béroul's interest in the Celtic sovereignty bride plot and his indulgent perspective with regard to the ill-fated lovers. At the same time, however, neither the lovers themselves nor the feudal world with which they contend would seem to have the privilege of recourse to such a rationale. The resultant ambiguity makes Subrenat wonder whether Béroul truly perceived a valid bond between the adulterous lovers, but he astutely identifies the nature of the competing moral frameworks:

> En tout état de cause, que l'on admette ou non les dernières conclusions que nous suggérions, la situation garde toute sa complexité puisque Marc et Yseut seraient époux aux yeux des hommes et de l'Eglise, mais non aux yeux de Dieu qui serait plutôt enclin à considérer Tristan et Yseut comme liés par l'amour. (232)
>
> [*In either case, whether or not one accepts these last conclusions, the situation remains no less complex since Mark and Isolde would be spouses in the eyes of the Church but not in the eyes of God, who instead would be inclined to see Tristan and Isolde as being bound by love.*]

In subtle contradistinction to the approach taken by Subrenat, we have tried to show how such pagan traditions could be seen to provide *tacit* authority for the emerging doctrine in light of very limited scriptural and canonical sources strictly defined. This perspective helps us to resolve the ambiguities and fatalism of the text by placing them within a typological, rather than explicitly doctrinal, discursive strategy. Ostensibly, the world of *Tristan* is *not* the contemporary world of Béroul and his audience although it offers a veiled critique of traditional marriage practices and an implicit defense of emerging doctrines. Unlike the antique world of Paris and Eneas, however, the world of Tristan and Iseut is not overtly pagan either. As in *Girart de Roussillon*, for example, the troubled protagonists find themselves at odds not only with feudal hierarchies and political forces but in conflict with figures representing traditional Church authority, as Subrenat has suggested.

We have also pointed out that emerging marriage doctrine was fostered and influenced in part by larger shifts in ethical thinking that placed a new emphasis on subjective intent over factual and objective reality (Chapter Three). Hunt's analysis complements Subrenat's approach by applying the new principles of truth and sin to his reading of Béroul's *Tristan* and the episode of the "false oath" (p. 515: "Béroul is guilty less of special pleading than of exposing the hollowness of conventional appearances as a touchstone of morality; the weakness, in short, of the "objectivist" doctrines against which Abelard was reacting"). Hunt relates the new emphasis on subjective intent to the poem's feudal critique in ways that support our analysis of affective intent in the Oxford *Girart de Roussillon*. The bad barons, for example, act according to the prerogatives of objective law in condemning Tristan and Iseut, but their intentions are obviously motivated as well by self-interest, envy, and malice. As paragons of sincere love and "good faith," on the other hand, the lovers (despite their legal transgression) maintain a superior ethical posture and (as feudal subjects) serve the king faithfully in all other respects (Hunt 529–30; 534–5).

In what we have discussed so far there is little contradiction with regard to Béroul's own explicit perspective. Throughout the text Béroul makes reference to the feudal virtues of Tristan and to the treachery of the barons who feign to act in the interest of their royal overlord. In addition, the figure of Husdent (the lovers' loyal canine companion) emblematizes the positive value attached to subjective honor and loyalty within a feudal structure. I believe, however, that emerging marriage doctrine has a significance that goes beyond an attenuated judgment of the apparent marital transgression in and of itself and which accounts for the very importance of the adultery plot to begin with. Hunt contends that the principle of consent had a specific meaning in contemporary ethical thought as well. In this context, consent referred to the notion of subjective intention and conscience (as opposed to free choice). According to Hunt, therefore, the potion's coercive effect excludes culpability to the very extent that it excludes consent (530–1). But to see the potion as a coercive element and discrete narrative device is to obscure its mystical significance. Hunt's logic also does not explain Béroul's depiction of the lovers' encounter with the hermit Ogrin as the effect of the potion declines. Despite Ogrin's exhortations, the lovers do not genuinely repent nor do they desist. Le Gentil reminds us that the lovers never prove to be totally oblivious to the consequences of their actions in relation to King Mark ("Tristan" 115). The love potion, therefore, is "avant tout symbole poétique, il devait servir à peindre plutôt qu'à absoudre" ("Tristan" 115). Kay has proposed that the encounter with Ogrin serves to redeem the lovers morally prior to the swearing of the false oath, thereby reducing the subjective moral liability of this latter transgression. But as Kay herself intuits, Béroul seems intent on denying us such a straightforward accounting of the lovers' actions (Kay 192–3).

In the *Tristan* story, as in *Girart de Roussillon*, we must realize that the dispensation binding the lovers escapes not only objective and customary laws but also conventional moral and spiritual teaching. Whatever contemporary ethical philosophy may influence Béroul and provide a gloss for his narrative, we should also keep sight of the immediate drama he depicts: Tristan and Iseut are not only marginalized with regard to the feudal court, but with regard to strict moral teaching as well. Their exchange with Ogrin is not depicted as supporting the conventional dichotomy between willful indulgence and moral redemption, nor does it reflect a conventional transition from forgetfulness to repentance. In the first encounter with Ogrin, the lovers acknowledge that they are aware of the spell put upon them by the potion but in no way follow Ogrin's lead by expressing repugnance or regret for their plight:

> Ogrins li dist: "Et quel confort
> Puet on doner a home mort?
> Assez est mort qui longuement
> Gist en pechié, s'il ne repent.
> Doner ne puet nus penitance
> A pecheor; souz penitance!
> L'ermite Ogrins molt les sarmone,
> Du repentir consel lor done. . . .
> A Tristan dist par gran desroi:
> "Que feras-tu? Conselle toi.
> — Sire, j'am Yseut a mervelle,
> Si que n'en dor ne ne somelle.
> De tot an est li consel pris:
> Mex aime o li estre mendis
> Et vivre d'erbes et de glan
> Q'avoir le reigne au roi Otran.
> De lié laisier parler ne ruis,
> Certes, quar faire ne le puis." (vv. 1387–94; 1399–1408)

[*Ogrin says to them: "And what comfort can one offer to a dead man? All but dead the one who for a long time lives in sin, should he not repent. No one can offer a sinner repentance where there is not first some penitence!" The hermit Ogrin preaches to them at great length, advising them to repent. . . . Exasperated, Orgin says to Tristan: "What will you do? Think on it." —"My lord, I am deeply enamored of Isolde, such that I can find no rest and no sleep. My mind is made up entirely: I prefer to be a pauper with her and live off roots and acorns than to rule over the marvelous kingdom of Otran. Make no mention of my leaving her, nor ask me to do something I cannot do."*]

Of course, both Tristan and Iseut blame the potion for their plight (vv. 1384 and 1412), but neither their awareness of the potion's effect nor Ogrin's moral admonishments diminish the mutual affirmation of amorous fidelity. We might say that Ogrin and the lovers enact a sort of "dialogue de sourds:"

> Ogrins li dit molt bonement:
> "Par foi! Tristran, qui se repent
> Deu du pechié li fait pardon
> Par foi et par confession."
> Tristran li dit: "Sire, par foi,
> Q'el m'aime en bone foi,
> Vos n'entendez pas la raison:
> Q'el m'aime, c'est par la poison.
> Ge ne me pus de lié partir,
> N'ele de moi, n'en quier mentir." (vv. 1377–1386)

[*Ogrin says to him: "In faith, Tristan! God pardons the sins of one who repents in words and in good faith." Tristan replies: "Sire, in faith, you fail to understand the reason why she loves me in good faith. If she loves me, it is because of the potion. I could not part from her, nor she from me; nor do I seek to deceive you in this."*]

In this exchange the lovers' sin is paradoxically cast as virtue. The language of the Church is adapted to a distinct but parallel plane of secular connotation. In fact, we find that the lovers' fate reveals itself in terms of the essential feudal virtue extolled in *Girart de Roussillon*, namely the virtue of *bone foi* explicitly aligned (as is it is in *Girart de Roussillon*) with its contrary *mentir*. Béroul also plays on the religious and vernacular connotations of "sin" (*pechiez*) when Iseut, distraught, says to Ogrin:

> Sire, por Deu omnipotent,
> Il ne m'aime pas, ne je lui,
> Fors par un herbé don't je bui
> Et il en but: ce fu pechiez.
> Por ce nos a li rois chaciez. (vv. 1412–16)[27]

[*Sire, in the name of almighty God, he does not love me, nor I him, except on account of the potion of which I partook and he partook. This was our sin / error. This is the reason why the king drove us away.*]

Here the ambiguity of the demonstrative pronoun ("Por *ce* . . .") only adds to the problems of cause, effect, volition, and action complicating moral judgment. The corruption originating with the potion resonates with the concept of original sin itself, as opposed to any remediable form of misconduct or indulgence.[28] The futility of this exchange leaves Ogrin skeptical and frustrated:

> Diva! [A Dieu va!] cil Dex qui fist le mont,
> Il vos donst voire repentance! (vv. 1418–19)

[*God help us! And may this God who made the world give you true repentance!*]

Ironically, Ogrin's outburst foreshadows the episode of the "false oath" during which God does, for all intents and purposes, exonerate the lovers. These exchanges with Ogrin, however, take place during an initial encounter which precedes the attenuation of the potion's effect. After three years, the effect wanes. Immediately, the lovers decide to seek counsel from the hermit once again. Should we not expect that this time Ogrin's admonishments lead to a meaningful act of repentance?

What Béroul delivers, in fact, is an ironic reversal of the penitential drama. Repentance involves an initial revelation or spiritual awakening to be followed by regrets and a new course of action. Béroul's ironic strategy must be understood in terms of the logical relation between both stages. On the surface, the lovers become aware of their transgressions and appeal to Ogrin who will help them work out a settlement with Mark that facilitates Iseut's return to the court after the lovers' separation. Ironically, however, the lovers' awakening is, above all, a remembrance of lost luxury and celebrity. As the potion wanes, the lovers do not truly repent their love, they repent only the resulting infractions made against otherwise complementary feudal obligations (vv. 2161ff.). Just as the awakening implies no real change in their affections, the impending separation actually prompts the lovers to reaffirm and validate the bonds that unite them. Amorous desire is not replaced by either revulsion, repentance, or even neutrality. Instead, what was inscribed in the potion as a mysterious and unconscious impulse becomes the invisible foundation for a more lucid but no less compelling and no less enduring affective bond. In fact, the failed repentance issues, ironically, in an unexpected moral transformation: the lovers, rather than repudiating their relationship, validate and sanctify it (albeit clandestinely) by virtue of a formal vow. The very elements of the ritual act bespeak its larger significance. For here again, as in *Girart de Roussillon*, we can point to a convergence of feudal and marital ideals. Indeed, the clandestine marital rite mirrors the ideal feudal relation between lord and vassal, in which the symbolic exchanges represent mutual consent and sincere bonds of good faith with regard to future service:

> Amis Tristran, j'ai un anel,
> Un jaspre vert a un seel.
> Beau sire, por l'amor de moi,
> Portez l'anel en vostre doi;
> Et s'il vos vient, sire, a corage
> Que me mandez rien par message,
> Tant vos dirai, ce sachiez bien,
> Certes, je n'en croiroie rien,
> Se cest anel, sire, ne voi.
> Mais, por defense de nul roi,
> Se voi l'anel, ne lairai mie,
> Ou soit savoir ou soit folie,
> Ne face con que il dira,
> Qui cest anel m'aportera,
> Por ce qu'il soit a nostre anor:
> Je vos pramet par fine amor.
> Amis, dorrez me vos tel don,
> Husdant le baut, par le landon?
> Et il respont: "La moie amie,
> Husdent vos doins par drüerie.
> — Sire, c'est la vostre merci.
> Qant du brachet m'avez seisi,
> Tenez l'anel, de gerredon.
> De son doi l'oste, met u son.
> Tristran en bese la roïne,
> Et ele lui, par la saisine. (vv. 2707–2732)

[*"Dear friend Tristan, I have a ring set with a green jasper. Handsome lord, as a sign of your love, wear the ring on your finger. And should you one day care to send to me with any message, I tell you now, let it be clear, if I fail to see this ring, my lord, I will give the message no credence. But, for the orders of no king will I ever fail—be it wisdom or folie—to do the bidding of the messenger who brings to me this ring, for it is a matter of our mutual honor. This I promised you, lord, in the name of courtly love. Dear friend and lord, will you give to me as a gift our hound Husdant?" And Tristan replies: "My dear friend, out of courteous love I give you Husdent." —"Lord, I thank you. In exchange for the hunting dog you have put into my possession, take here this ring as my gift." She takes it from her finger and places it on his. Tristan kisses the queen, and she kisses him in return, sealing the pledge.*]

Ultimately, Béroul's treatment of the "love-drink" motif is neither as forced or as clumsy as some critics have suggested. By maintaining vestiges of the fatal Celtic *geis* and then contrasting the symptoms of the marvelous spell with the consensual bond, the drama elaborates something of a sentimental archeology (*post facto* and fictional) of *fin'amor*. Iseut's speech and the bond it establishes evoke familiar features of "courtly love" and lend a conscious moral dimension (the verbal engagement) to the love relationship. But by making this exchange into an extension of an erotic desire that is initially defined according to the implicit nature of the Celtic *geis*—i.e., a magical, impulsive, involuntary and unrelenting form of *eros*—Béroul gives a new ontological and subconscious depth to an otherwise predominantly rhetorical, mannered, and cultivated expression of desire and affection. The violence and mystery of the *geis* is also transferred onto a sacramental plane as the solemn vow (which amounts to the ratification of a valid consensual marriage) brings the lovers' bond into conformity with emerging Christian doctrine.

The potion, therefore, does not truly provide a pivotal device (as Hunt argues) by which to exonerate the lovers morally. Indeed, what Béroul suggests is that the lovers' plight is to be defined not by the dichotomies of sin and repentance but by a mystical dispensation that integrates the secular subject (and ultimately secular society) to a higher will. Ultimately, however, the "amorous rule" is meant to be the basis for a reintegrated society and not for a mystical grail quest. As the effect of the magical potion wanes, the truth of its mystical power becomes, symbolically (like the wine that represents Christ's blood and binds believers), the basis for a secular imperative based on friendship, loyalty, and good faith. The transition from intoxication to sobriety does not, therefore, coincide with a transition from coercion to volition or from transgression to redemption, but with a transition from an instinctive drive or *natural law* to *divine law*—to an ideal based not on arbitrary conventions but deriving from a "natural law" that is universal and transcendent.

By patenting the existence of an authentic secular *ordo*, the emerging matrimonial doctrine created an important breach in the monolithic doctrine of patristic Christianity and created a discursive space within which to imagine and elaborate an edifying secular ideology sanctioned by God but distinct from the canonical virtues governing clerical and monastic orders.[29] We have also seen (especially in our analysis of the *Roman d'Enéas* and the *Roman de Troie*) that this transformation of ethical values required first and foremost a transformation in traditional patristic attitudes with regard to *eros*, the formation of amorous bonds, and the nature of amorous psychology. Amorous desire—traditionally associated with carnal demons in medieval patristic theology—would now constitute an essential feature (i.e., *affectio maritalis*) of the very institution intended to unite secular existence and historical destiny with providential designs.

In *Girart de Roussillon* the subjective and spiritual bonds integrating feudal society are also extolled over hierarchical agencies and structures, whether secular or religious. Since marriage plays a pivotal role as the foundation of larger social alliances, the validity and integrity of matrimonial ties should be warranted not by external forms of custom or coercion but by sincere sentiments of mutual consent and affection. In this specific way, the poem's vision of secular order corroborates a central tenet of emerging doctrine—that marriage is a sacrament whose proper union has one valid basis vouchsafed by God and designated as the singular institution ordering secular affairs. The Old French *Tristan* narratives, by contrast, elaborate a more fundamental aspect of the same doctrine: namely, the very notion of a sacred secular order defined by a transcendent form of *eros*. The Ogrin episode emphatically dramatizes the emergence of a new space that escapes the old patristic dichotomy between religious and secular existence. We recall that according to the Pauline axiom, conjugal life represents an existence in an attenuated state inferior to the virtues of celibacy and religious devotion. The exchange between Ogrin and Tristan, however, suggests a new dispensation in which the secular itinerary now conforms to a transcendent order of its own right, not subordinate to, but parallel with, the religious order and governed by the laws of *eros*. The Old French *Tristan* is notable in its attempts to integrate a wide array of profane features of secular existence into the hallowed itinerary of its protagonists. As Kay points out, "the *Tristan* is generically somewhat indeterminate. For example, the terms 'epic' and 'fabliau' as well as 'courtly' are widely applied to Béroul. Maybe the *Tristan's* elusiveness of 'point of view' is connected with this generic blend" (186).

Of course, Kay's observation obscures the fact that the itinerary of Tristan and Iseut, like that of Girart de Roussillon, is also indebted to plot motifs taken from contemporary hagiography. One might argue that the martyrdom of the legendary hero Roland provides a precedent for this kind of influence. But the Ogrin episode clearly involves a new dynamic. Roland is, at least ostensibly, a martyr because of his fulfillment of objective religious values: he selflessly defends Western Christendom from the infidel. As adulterous lovers, Tristan and Iseut transgress conventional religious doctrine, but their transgression and their failure to "repent" issues—paradoxically—in an exile whose itinerary is modeled on hagiographic ideals of fidelity and estheticism. Like religious devotees, the lovers are rendered immune to their hardships by the love which motivates them (1364–6). In another passage, the lovers are described as if martyrs ("Trois ans plainiers sofrirent peine, / Lor char pali et devint vaine. / Seignors, du vin de qoi il burent / Avez oï, por qoi il furent / En si grant paine lonctens mis" vv. 2131–35).[30]

Ulle Erika Lewes has found that Guillaume de Berneville's Old French version of the *Vie de St. Giles* (a twelfth-century work derived from a tenth-century *vita*) is connected to the court of Henry II Plantagenet and embellished with details and scenes from courtly life that make it comparable to contemporary romances. Likewise, Lewes observes the influence of the St. Gilles legend throughout the *Tristan* tradition.[31] In Thomas's version, the hagiographic analogy is highlighted by the spiritual "bliss" experienced by the exiled lovers. The Norwegian Saga provides evidence of Thomas's treatment of the exile period:

> Lorsque Tristan fut revenu à la cour du roi Marc tout heureux et joyeux, il n'y demeura pas longtemps que le roi n'eût à nouveau découvert le grand amour que Tristan et Yseut avaient l'un pour l'autre, tout comme auparavant. Le roi en fut très fâché et affecté, se refusa à supporter cela plus longtemps de leur part, et les bannit alors tous les deux. Cela leur apparut cependant comme une chance, et ils s'en allèrent alors dans un grand désert. Mais ils ne réfléchirent guère à qui les pourvoirait en vin et en provisions, car Dieu voudrait sans doute bien leur procurer quelque norriture en quelque lieu qu'ils se trouvent. Et ils appréciaient beaucoup d'être tous les deux seuls ensemble. De tout ce qui existait dans le monde, ils ne désiraient rien de plus que ce qu'ils avaient alors, car ils avaient ce qui plaisait à leur coeur s'ils pouvaient toujours demeurer ainsi ensemble sans être accusés, et savourer leur amour dans le bonheur. [. . . .] Cette vie leur apportait beaucoup de gaieté et de plaisir parce qu'ils connaissaient nuit et jour la joie et le réconfort. (613)[32]

> [*Once Tristan had returned to King Mark's court, happy and delighted, he remained there only a short time before the king once again discovered the strong love that Tristan and Isolde had for one another, just as before. The king, hurt and very angry, refused to tolerate their conduct any longer, and so he banished both of them from the court. They saw this, however, as a lucky opportunity, and so the two of them went away into a vast desert. They hardly thought about who would supply them with wine and food, for no doubt God would take care to procure some amount of food wherever they might find themselves. And they were very grateful to be together by themselves. Whatever the world might offer, there was nothing they desired beyond what they presently possessed; for they had their hearts' content provided they could always remain thus together, free from blame,*

and able to savor their love in happiness. . . . This life brought them great
amusement and pleasure, because day and night knew joy and comfort.]

Although hagiographic motifs find their way into numerous Old French
secular narratives, Lewes points out that the *Tristan* narrative shows an
essential kinship with the structure of the St. Gilles legend. In the tradition
deriving from Thomas, for example, regret is expressed for the lovers' lost
happiness upon their discovery in the cave: "Such an attitude is, of course,
fundamentally antithetical to the principles of courtly romance, in which
reintegration into the royal court usually constitutes the climax of the
entire story" (48). In Béroul, the paradoxical transfers of ideology make sin
into virtue and virtue into sin. The fact that the lovers never successfully
reintegrate with the royal court seems to underline the fact that marriages
by mutual consent, as opposed to customary aristocratic marriages, provide the very *sine qua non* of an ideal feudal order, their absence resulting
in fragmentation and strife. We saw this same kind of strife identified by
Thomas's extended commentary on the suffering caused by invalid marriage constraints.[33]

After the *romans d'antiquité*, the Old French *Tristan* marks an even
more radical moment in which emphasis is no longer placed on the dynastic and political value of marriage strategies but on the mystifying and insuperable authenticity of the erotic bond itself as a supreme value in determining the subjective experience of the secular hero. This aspect of the Old
French *Tristan* tradition links the blossoming of Old French secular narrative to a larger Western tradition in which the cultivated and mannered
modalities of refined love are ontologically linked to more primitive
impulses which ratify their essential role in the evolution of cultural
supremacy (*translatio imperii*). In this way cultivated love is not restricted
to mere forms of poetic play and subjective conceit as it often is in the troubadour tradition for example. This vein of twelfth-century love builds on
the Ovidian ethos of civilized refinement while rejecting the cynicism of
Ovidian psychology. In conformity with progressive marriage doctrine,
therefore, the Old French *Tristan* tale defines an authentic form of natural
love which transcends sinful cupidity and has its place within a larger providential order although circumscribed by purely secular forms of consciousness and experience. This not only invites the importation of hagio-
graphical plot motifs into secular narratives, but reflects the emergence
within medieval culture of an independent sphere of secular discourse and
values on a plane distinct from but often analogous to traditional patristic
and monastic planes of moral discourse.

MARRIAGE DOCTRINE AND THE SECULAR ORDER: *PARTONOPEUS DE BLOIS*

Secular and religious themes converge in a similar way in the marital drama
depicted in *Partonopeus de Blois*. We have already seen, for example, that

the supernatural associations surrounding Mélior also serve to bring to the fore larger questions concerning the anarchic and violent aspects of erotic love that underlie consensual models of courtship and marriage. The poet first allays some of these fears by transferring the demonic tropes onto his critique of aristocratic marriage strategies, in which entrenched interests inevitably result in their own forms of deceptive coercion. By evoking the specter of the pagan fairy figure, however, the poet also raises fundamental concerns about erotic love in relation to traditional forms of Christian morality. Earlier, we saw that these kind of fears increased the efforts by Partonopeus's mother to separate her son from Mélior. Actually, Partonopeus himself is duly suspicious of Mélior's nature. But Mélior insistently defends herself against such suspicions as the poet begins to reorient the pagan fairy figure in a manner consistent with orthodox Christian faith:

> Mais je sai bien que vos cremés
> Que jo ne soie aucuns maufés
> Qui tant vos face par losenge
> Qu'en aucun mal pechié vos prenge
> Por faire vostre arme [âme] périr;
> Mais ne vos voel de ce servir.
> Je croi en Deu le Fil Marie,
> Qui nos raienst de mort à vie,
> Et por lui pri que vos m'amés;
> Se por el faire nel volés,
> Rien ne m'orés dire ne faire
> Qui onques li soit à contraire. (vv. 1529–40)

> [*But I know for sure that you fear that I might be some demon who so flatters and deceives you that I draw you into some evil sin in order to imperil your soul. But I have no intention of betraying you. I believe in God, son of Mary, He who delivers us unto life from death. And for Him I ask that you love me. And if you do not wish to do so on His account, you will never hear me say or do anything unpleasing to His will.*]

Of course, Mélior, like Marie de France's pagan bird divinity in the *lais* of "Yonec," is "Christianized" but not entirely demystified. Instead the mystical aspect of the pagan and erotic elements—as well as the secular love affair itself—all take on a Christian aspect.

This reorientation of the figure of Mélior is part of a larger attempt to define a new secular ethos and model of marriage which involves the intersection of Christian and erotic discourses. This strategy becomes clear in the final contest between Partonopeus and the sultan for Mélior's hand. In this context, Partonopeus's martial and amorous conquest becomes the emblem of cultural superiority and Christian hegemony. Thus, the marriage conquest once again converges with a larger historical *translatio imperii.* But this triumph is defined not only as the triumph of innate nobility and chivalric refinement—exemplified by Partonopeus—

but in broader terms as the triumph of a Christian model of courtship and marriage.

By means of Partonopeus's contest with the pagan sultan, the poet evokes a patristic morality inhospitable to the yearnings and indulgences of erotic desire and amorous affections. After Partonopeus's marriage, the sultan—still ravished by an enduring love for Mélior and seeking revenge against Partonopeus—rails against the Christian godhead in an epistle where he laments his amorous suffering. The sultan's letter, addressed to Mélior, reminds us that the West is indebted to pagan traditions for the cultivation of a divine concept of erotic passion typically excluded from conventional Christian categories of divinity:

Diex, fait il, quel diu requerrai?
Lequel haut roi deprierai?
Li nostre dieu seulent amer,
Chascuns d'euls seult avoir sa per:
Phebus et Mars et Jupiter,
Neïs Pluto li diex d'enfer.
Nuls de nos diex n'est si pervers
N'ait amor donnee en travers.
Se Melior fust sarrazine,
Assés demandasse mechine:
Cil ont en lor amour apris
Con doivent aidier as chetis.
Mais ele tient une autre foi,
Qu'ele a une nouvele loi
D'un Diu qui ainme chaasté,
Qui onques en tout son aé
N'ama d'amor, tant par est fiers.
Il n'est pas Diex as chevaliers,
N'as jones gens, n'a dames beles,
N'a envoisies damoiseles.
Il n'a mie de dosnoi cure,
Ains ainme une gent forte et dure,
Uns angoisseus jeüneours,
De lor cors uns justiceours,
Qui n'aiment n'ami ne parent,
Ains donnent le lor por noient
Et n'aient cure de delit;
Et cest siecle a tant en despit
Por un autre qu'il lor pramet,
don't il si lonc respit lor met,
Que ja por tant com il sont vif
N'en auront joie li chetif:
Aprés la mort lor a promis
Ques metra en son paradis. (T 141 b, vv. 1–34)[14]

[*"Oh gods,"* he said, *"Which god shall I beseech? Our gods are accustomed to love; each one of them has a paramour—Phoebus, Mars, and Jupiter, even Pluto, the god of the underworld. None of our gods is so eccentric; none has so disparaged love. If Melior had been a Saracen, enough to ask the girl for satisfaction—after all, her people have learned a love that teaches mercy for those in need—but, alas, she is of another faith, just as she adheres to a new covenant from a God who favors chastity, who never in all His life loved with passion so stern and proud is He. He is not a God suited to knights, young bachelors, beautiful ladies, or cheerful damsels. He cares nothing for amorous flirtation, adoring instead a stern and stoic race, perennially worrying, fasting, and passing judgment on a people who love neither friend nor kin, thus loving in vain and having no interest in amusement. With scorn He holds this world inferior to another that He promises them and which He withholds from them so long that never, as long as they may live, will they take any joy in it, these miserable disciples—only after death does He will put them in His paradise."]*

In this speech, as in the creative exploitation of the demonic fairy motif, the author of *Partonopeus de Blois* once again confronts the ambivalence inherent to the emerging convergence of Christian and secular views of love and marriage. Just as the very notion of marriage based on mutual consent and affection invites concerns over a subversive and dominant female animus, so too does it evoke an image of man not made after a chaste and austere monotheistic God, but according to the image of classical pagan gods. How then does the author of *Partonopeus de Blois* define this new secular space?

For one thing, the figures of Partonopeus and Mélior are assigned a normative Christian standing while the Persian sultan Margaris and Partonopeus's companion Anseau (converted from his pagan beliefs but baptized in an Albigensian chapel) serve as foils.[35] Together, the character profiles contribute to the poet's revisionist subtext. Only Partonopeus's amorous fortunes will take the form of a happy marital resolution, but all three figures share a noble aspect based on the vitality of their amorous passions and on their demonstrations of amorous fidelity. At the same time, the religious distinctions seem to have some relation to their respective amorous destinies. The sultan Margaris, for example, argues that no one of his gods (actually classical gods!) "n'ait amor donnee en travers" as the Christian god seems to have done by making Mélior resistant to his pleas. Yet it is Margaris who is insistently courting a bride of a different faith and challenging the private, public, and divine (it would seem) wills that have sanctioned the marriage of Partonopeus and Mélior. In addition, the marriage between Partonopeus and Mélior has now been consummated as well. In asking that Mélior respond to his pleas and "remedy" his sufferings, the Saracen leader Margaris is not simply courting his beloved but encouraging an adulterous transgression. This is implied by Mélior's own response, when she affirms that she is a *consenting* party to her present marriage and has no intention of "sharing" herself:

Li soudans est de mout grant pris,
De valour et de courtoisie
Assés vault, et mout le mercie
De s'amour qu'il m'a presentee
Et de cuer et de cors donnee.
Mais se il veult avoir m'amour,
Ce li porrés dire au retour:
Si s'en revoist en son païs,
Car ne voil avoir .II. amis.
Li miens si est Partenopex,
Icelui me gart li vrais Diex! [T 145 a, vv. 4–14][16]

[*The Sultan is a man of great merit, a man of valor and courtesy, worthy
enough, and much do I thank him for the love he sends, offering a love of
heart and body. But if he wants to have my love you can tell him this for
me: better he should go back to his country, for I do no wish to have two
lovers; my lover is Partonopeus—may the true God protect him!*]

Thus, there is also a false note in the sultan's implication that Mélior is simply unwilling to respond to a noble lover and to give up her chastity
("Qu'ele a une nouvele loi / D'un Diu qui ainme chaasté . . .).

The reference to chastity, however, evokes what is in fact a secondary
thematic motif in the poem. For the *Partonopeus* poet takes a quite out-
spoken stance on this question. In fact, one can say that the poet advertis-
es his own membership within the amorous fraternity we have identified.
Like the sultan, he disparages chastity in women (especially in beautiful
women!) and sees this supposed virtue not only as a waste but also as contrary to Beauty, which is inclined to nobility, generosity, and gaiety (vv.
6217ff.). Like the sultan, he also suffers from unrequited love, from the
austere rejections of a chaste woman (vv. 6255–64). Therefore, although
the narrative perspective departs from that of the sultan, it does not view
secular morality according to strict patristic ideas about sex, chastity, virginity, and abstention. By evoking, through the device of the sultan's
prayer, a contrast between Classical divinities and an austere Christian godhead, the poet of *Partonopeus de Blois* defines a revisionist morality integrating Classical attitudes toward *eros* into a Christian ethic of marriage
based on affection and spousal consent (whose prerogatives, so defined,
attenuate harsh condemnations of adultery and reorient patristic views on
the sanctity of chastity in women).

According to a similar logic, the poet also seems to treat the more austere tenets of *fin'amor* with some reservation. Early in the narrative,
Partonopeus is befriended by Anseau at the moment when he himself is
contemplating suicide for having betrayed Mélior's trust and having lost
her good favor. Anseau, somewhat like Partonopeus, is a figure of disenfranchised nobility. He also has a marginal status with regard to religious
orthodoxy, having converted from his pagan faith as an adult and having
been forced to consummate the urgently sought conversion in an

Albigensian chapel (vv. 5659–68). Like his marginal religious status, his status as a lover also reflects a certain marginality in contrast to Partonopeus. He is obsessed by memories of a beloved named Auglaire, yet we learn that he consistently rejected her advances and committed himself to his unrequited passion due to his inferior birthright and out of gratitude for the monarch who helped him (Auglaire is the niece of the German Emperor, while Anseau's birthright is obscure). When Auglaire tries to "take" him, he flees—in direct contrast to Partonopeus who aggressively "takes" Mélior during their first night together). His refusals earn him the nickname "l'Oublieux," so that the connotations of moral neglect (*s'oublier*) are now attached not to an act of sexual indulgence but to an excessive form of amorous restraint and abstention.

In yet another diatribe against chastity we hear further echoes of the sultan's critique of the Christian faith, a critique targeting women who seek reclusion and chastity. The poet (describing the preparations for the tournament) nostalgically regrets that refined love and dalliance have been supplanted by religious austerity and that contemporary women remain secluded in church and fixed on their prayers (vv. 7985–8036). Of course the poet had already displayed a rebellious revisionist spirit at the outset of the narrative. In weaving a dynastic myth which links the early Franks and by extension the hero Partonopeus with the Trojan line of Paris and Hector, the poet confronts a well-known apocryphal account of the Troy story which involves a Trojan traitor (here the figure is Æneas's father Anchises) precipitating the fall of Troy. Therefore, to maintain the integrity of the Trojan line, the poet suggests that Æneas is actually born of a different father. In reference to this historical anecdote, he declares:

> Miols vaut bons filz à pièces [i.e., a pechiez] nez
> Que mauvais d'espouse engenrés. vv. 313–14
>
> [*A good son born in sin is worth more than a bad one conceived in wedlock.*]

Hence, the poet clearly takes a provocative stance in relation to orthodox Christian morality. Nonetheless, this same poet exploits religious motifs throughout the narrative as a way of linking Partonopeus's ostensibly secular itinerary with new theological attitudes toward marriage.

IRONY AND THE *VOX POPULI: CLIGÈS* AND *L'ESCOUFLE*

If we are to take the narrative perspective in *Partonopeus de Blois* for what it is, then we are forced to acknowledge that twelfth-century society, whatever its prejudices and dogmas and however severe its moral and customary laws, contained an intellectual tradition that can be qualified as "humanistic" by virtue of its interest in heroism, eroticism, psychology and sentimentality. But to so boldly condone adultery and to scorn religious devotion and chastity—is the narrative persona to be taken at face value;

is there not an *obvious* irony in the author's irreverent viewpoint? If one were to apply a strict standard of religious teaching, then obvious irony there is. But what we have tried to emphasize in our study of marriage fictions is that the *roman* is a genre that posits a distinct standard. From this perspective, religious devotion is not *necessarily* aberrant, but for women who live in a temporal court and as members of the secular order to define virtue in terms of chastity, virginity, and seclusion *is* an aberration. Within the parameters of emerging doctrine, therefore, we find principles that allow us to read tales of secular eroticism and adultery in a positive light. However, the problem of authorial or ironic "intent" will not be so simply resolved. For either the new doctrine itself, controversies surrounding it, or various interpretations and extrapolations of it, could become the target of irony in the case of any given author. I have already suggested, for example, that Chrétien de Troyes seems to use irony in order to maintain a critical stance with regard to the dangers of unchecked libidinous desire. From a clerical perspective, it is perhaps inevitable that the *comédie humaine* of secular life will invite ironic jesting; does this mean that the very treatment of secular courtship and adultery in twelfth-century literature is to be characterized as inherently ironic?

If the twelfth-century *roman* does indeed define a secular moral arena relatively independent from orthodox Church morality, then the conventional evocation of Christian motifs with regard to secular love, whether by literary forms of allusion or analogy, need not be appreciated in strictly rhetorical terms—as a form of poetic hyperbole, for example—nor in terms of a reductive ironic viewpoint. Instead, the religious analogy reveals the extent to which the secular society and subjective experience (i.e., of the secular subject) conform to transcendent laws of Christian grace and dispensation which are akin to those which govern monastic life and traditional forms of religious experience. Having defined this new perspective through our examination of *Girart de Roussillon, Tristan et Iseut,* and *Partonopeus de Blois,* I would now like to turn very briefly to Chrétien de Troyes's *Cligès* and to Jean Renart's *L'Escoufle,* for the reason that these works contain complex uses of irony which elicit conflicting responses and "ironic" readings.[37]

In Thomas's *Tristan,* it is hard to judge at times whether to interpret the hagiographic parallels in terms of an idealization of divine *eros* or in terms of the kind of ironic perspective championed by Robertson. In *Partonopeus de Blois,* on the other hand, we see the author attempting to define a model of amorous desire and consent sanctified in marriage and distinct from "unorthodox" tendencies. The figures associated with amorous "heresy" are so identified by a parallel association with religious heresy. In *Cligès* one also finds multiple models and perspectives. However, Chrétien's use of comic irony makes the reader's task more complex. Although Chrétien appears clearly to condone a model of marriage based on natural reason, love, and consent, he condemns reclusive and clandestine forms of marriage and courtship.

In *Cligès*, as in the *Æneid*, marital destiny remains tied to a larger social, historical, and political order. It is interesting, therefore, that Chrétien de Troyes also takes up another element from the Classical and medieval Dido episodes, namely the device of the *vox populi*. We have already remarked on the use of this device in the *Roman d'Enéas* and the *Roman de Troie*. In his comments on Béroul's *Tristan et Iseut*, Subrenat implicitly points to the same development of dynamic moral perceptions and judgments when he asserts that the marriage between Mark and Iseut is one thing "aux yeux des hommes et de l'Eglise" and another "aux yeux de Dieu." In our own analysis of the Ogrin episode, we saw just how clearly Béroul depicted the gap between traditional religious and penitential viewpoints, on the one hand, and the experience of the two lovers, on the other. In short, this device explains both the moral ambiguities that surround critical assessments of sacred and profane discourse in the secular *roman*, while also providing a key to their resolution. The key, of course, is our understanding the dynamic gap between customary attitudes and perceptions, on the one hand, and higher ethical laws, on the other.

In the works we have studied, the representation of clandestine love affairs will typically solicit a normative social judgment like that expressed through the *vox populi* device. The normative viewpoint sets in relief a higher ethical law and providence. Gautier's *Eracle* provides an additional case in point. Here, as compared with other examples we have touched upon, the device is more directly aligned with the wisdom motif itself—for it is the prophet and sage figure Eracle himself who laments the common perceptions that will mar his own reputation. At the narrative juncture in question, the Emperor Laïs has just discovered that he has been betrayed by his wife Athanaïs during his absence from the court. Eracle is in the process of demonstrating to the emperor that it is he (the emperor) who wronged Athenaïs and that the honorable thing to do is to release her from her vows, given her love for Paridès (vv. 4961ff). The sage points out that Laïs drove her to commit adultery by betraying his own faith in the empress and condemning her to a severe sequestration. Under the actual circumstances it will, nonetheless, still *appear* to the feudal populace, that Athanaïs is a woman of tainted character, that Laïs is the unfortunate victim of betrayal, and that Eracle himself erred in his initial judgment of Athenaïs's moral purity:

> Tort ai, ne l'a pas desservi,
> mais vostre cors que je mar vi,
> c'on dira cent ans ci après :
> "Cil qui se feme tint si pres
> faussa Eracle son devin ;
> la dame prist malvaise fin." (vv. 5017–5022)

[I get the blame, but I do not deserve this; curse the day I saw you [your court?], for a hundred years hereafter it will be said "The one who guarded his wife so closely was misled by his sage Eracle; this lady came to a bad end."]

Thus, while offering a logic which rather blatantly justifies adultery as a response to spousal coercion, the sage figure posits the operation of a sophisticated moral law, one that transcends customary ethical judgment and clashes with popular attitudes and perceptions. Gautier's text is unique in its extended elaboration of the *vox populi* topos, and in its attempt to situate it within a larger and explicit scheme of psychological and moral teaching:

Ne diront pas que je vos dis
que bone eüst esté toudis
s'ele ne fust emprisonee ;
car toute gens s'est adonee
et a mal dire et a mal faire,
ja nes orrés un bien retraire ;
mius volent mal dire et mentir
que nule rien bien consentir
que on le die de nului. (vv. 5023–31)

[They will not report that I warned you that she would have been virtuous forever if she had not been imprisoned, for it is the natural disposition of all people to speak ill and do harm. Never will you hear from them a favorable account of the matter. Rather they are happier to speak slander and lies than to acknowledge any good reported about anyone.]

The universal accusation reminds us at first of the subjective pride of the lyric subject who rails against the ubiquitous *losengiers*. In the remainder of the speech, however, Gautier delivers a more subtle psychological analysis of the phenomenon personified in Virgil as winged *Fama*:

Cascuns se sent si plains d'anui
ne velt pas c'on des autres die
ce qu'il en soi ne cuide mie.
Puet estre uns peciés les deçoit,
que nus son mehaing n'apeçoit
et voient es preudomes l'ombre
de cele riens qui les encombre.
Ne voient pas don't l'ombre vient
qui si tres pres des cuers lors tient.
Ne mais que calt ! li biens vaintra,
et aucuns preudom le dira :
"Certes s'Eracles fust creüs,
Laïs ne fust ja decheüs." (5032–44)

[Each individual feels so full of pain and spite that he or she does not want people to say things of others that they do not recognize in themselves. It might be that some error or sin deludes them such that not one of them

*perceives his or her own defects. Then they project upon people of wis-
dom that very thing which burdens them. They do not see where this
shadow comes which in fact clings closely to their very own heart. Come
what may! Virtue will win out, and some person of integrity will say to
them: "In truth if Eracle's advice had been followed, Lais would not have
ended in ruin."*]

Gautier's text makes it clear that the moral paradoxes surrounding representations of marriage and adultery in the twelfth–century *roman*—paradoxes that continue to trouble modern critics—could not be explained
according to black and white applications of legal or patristic precedent.
Indeed, Eracle's speech warns the reader that the legal and moral questions
raised by emerging doctrine run counter to habitual attitudes. The fictional marriage drama anticipates the very controversies that will surround the
tale's reception—in its own day, a "hundred years later," and today.

In the end, however, Gautier's text—while it provides a relatively sophisticated psychological analysis of a common motif—immediately resolves
any ideological ambiguity by virtue of the protagonist's status as an authentic sage figure. Nevertheless, the explicit preoccupation with the motif is an
indicator of its literary currency. In *Cligès*, for example, Chrétien will bring
an ironic twist to the same topos, such that it illuminates the moral and
psychological limitations of the very figures who are ostensibly the victims
of distorted perceptions. Fénice, for example, does not express remorse
because of her adulterous desires. Rather, it is public opinion, the *vox populi*, that troubles her. Nor is this simply a convenient hypocrisy (or so it
seems, at least). For, as Fénice points out, she does not *privately* consider
herself a married woman:

> Et sachiez bien, se Dex me gart,
> Qu'ainz vostre oncles n'ot en moi part,
> Car moi ne plot, ne lui ne lut.
> Onques ancor ne me conut
> Si com Adanz conut sa fame. (vv. 5219-23)

> [*And let it be known, God defend me, that at that time your uncle had no
> claim to me (no intimacy with me) for it was not my wish. Never to this
> day has he known me in the way that Adam knew his wife.*]

At the same time, however, she is the titular wife of Alis—and this defines
her public persona:

> A tort sui apelee dame,
> Mes bien sai, qui dame m'apele
> Ne set que je soie pucele. (vv.5224-6)

> [*Wrongly am I called a lady, but I know well that anyone who calls me a
> lady does not know that I am a young (virgin) girl.*]

The affinity between the lovers and the well-known Tristan and Iseut is
here apparent. However, in a conversation the preceding day with Cligès,

Fénice suggested that Tristan and Iseut merit repudiation:

> Vostre est mes cuers, vostre est mes cors,
> Ne ja nus par mon essanplaire
> N'aprendra vilenie a faire,
> Car quant mes cuers an vos se mist,
> Le cors vos dona et promist
> Si qu'autres ja part n'i avra. . . .
> Se je vos aim et vos m'amez,
> Ja n'en seroiz Tristanz clamez,
> Ne je n'an serai ja Yseuz,
> Car puis ne seroit l'amors preuz,
> Qu'il i avroit blasme ne vice. (vv. 5234–9; 5243–7)

> [*Your heart is mine, your body is mine. No one will ever learn to commit
> a vile wrong because of the example I am setting. For when my heart
> placed itself in you, it also gave and promised you the body such that
> another would never have any part of me . . . . If I love you and you love
> me, never will people say you are Tristan, never will I be Isolde. For then
> love would not be noble if it were mixed with any vice or blame.*]

Fénice's contention, as Nelson has pointed out (81–2), is that Iseut and
Tristan had a tainted affair because the queen simultaneously maintained
relations with King Mark (vv. 5244–5251). During the exchange on the following day, therefore, Cligès follows step and proposes they flee together
to Brittany, thus distancing Fénice from the emperor. But Fénice rejects the
idea, once again evoking the tragic fate of their doomed Celtic counterparts. This time, however, the hesitation no longer involves the question of
her sharing her body with two men. Instead, it is public perception specifically—the *vox populi*—which concerns her:

> Ja avoec vos ensi n'irai,
> Car lors seroit par tot le monde
> Ausi come d'Ysolt la blonde
> Et de Tristant de nos parlé.
> Qant nos an serïens alé
> Et ci et la totes et tuit
> Blasmeroient nostre deduit.
> Nus ne diroit ne devroit croirre
> La chose si com ele est voire. (vv. 5294–5302)

> [*Never would I run off with you in this way, for then people the world
> over would speak about us as people once spoke about Tristan and Isolde.
> Once we will have departed, everyone everywhere will speak ill of our dal-
> liance. No one can be expected to believe, nor will they report the thing
> such as it is in truth.*]

As Fénice continues her speech, she herself will indeed evoke a higher
law, that of the New Testament. This is where Chrétien introduces an original and ironic twist. Instead of contrasting the public perception with the
imperatives of her own moral conscience as it is seen by the Divine Judge,

she equates moral character with public perception—at least this is the implicit logic behind her distorted reference to the words of St. Paul:

> Mes le comandemant saint Pol
> Fet boen garder et retenir:
> Qui chaste ne se vialt tenir,
> Sainz Pos a feire bien anseingne
> Si sagement que il n'an preingne
> Ne cri ne blasme ne reproche. (vv. 5308–13)

> [*But it is well to remember closely the words of St. Paul: to the one who does not wish to remain chaste, well does St. Paul teach to act so wisely (cleverly) that he or she incur neither accusation, blame, nor reproach.*]

Fénice gives a strictly amoral meaning to the word *sagement*; she means *astucieusement*. Her reference to Paul's text serves to introduce the elaborate scheme she has concocted so as cleverly to deceive the public and thereby escape public accusations (cf. vv. 5314ff.: "Boen estoper fet male boche . . . "). Hence, instead of seeking consolation in the fact that her love is innocent and valid "aux yeux de Dieu," Fénice will attempt to construct a moral self by manipulating public perception and taking refuge in clandestine seclusion.

The ironies remind the reader that marriage plays both a public and private role, while they caution (at least implicitly) against separating the two. Fénice's dilemma clearly requires a resolution of the conflict between her private desires and public status, but she errs by virtue of a cultural orientation that is purely secular. She fails to see the consensual principle that makes her desire morally defensible. Seeking to protect her external "honor," she confuses her public identity with her moral identity. Hence the ironic angle derives not from an implicit contrast between secular conduct and patristic morality, but from the protagonist's attempt to establish moral credibility in terms of public perception—this within a narrative genre that implicitly subordinates customary law and public moral judgment to a transcendent ethical will based on natural law and harmonized with emerging theological doctrines on marriage. Thus, Chrétien is sympathetic to Fénice's plight and favors a consensual model of marriage. At the same time, however, he uses Fénice's own limited perceptions and flawed course of action to introduce a competing model, in which an albeit compelling and natural bond of love is perpetuated in a totally clandestine and unnatural manner. In both *Cligès* and *Partonopeus de Blois*, therefore, distinctions are made between competing forms of love, but a consensual model involving amorous courtship prevails nonetheless. Chrétien's use of narrative irony notwithstanding, any strict Robertsonian opposition between secular and patristic values fails to obtain.

If we turn to two later works in our corpus, both by Jean Renart, we find an idealized view of amorous courtship where the shadow elements are clearly those that obstruct the destiny of the amorous heroes. Both

L'Escoufle and *Guillaume de Dole* exemplify the extent to which amorous sentimentality comes to represent the cardinal virtue in a broader ideology of cultural refinement and secular dominion.

In the case of *L'Escoufle* in particular, we find a salient example of a secular drama centering on an amorous adventure and an idyllic vision of love which borrows heavily from hagiographic narratives. The juxtaposition of secular and religious elements is perhaps most notable in the scene describing the arrival of Duke Richard (Guillaume's father) in the holy city of Jerusalem (vv. 460ff.). Interestingly, this scene—featuring a golden chalice inscribed with images from the tale of *Tristan and Iseut*—has, like the tale itself, been appreciated above all for its apparently ironic portrayal of secular love.

After his arrival in Jerusalem, Count Richard is escorted by the King of Jerusalem to the Temple that shelters the Holy Sepulcher. As part of the religious and civic pageantry, Count Richard presents, in the form of an offering to be placed at the holy altar, a valuable gold goblet with enameled images depicting the amorous adventures of Tristan and Iseut. After ennumerating the various episodes of the *Tristan* story evoked by the engravings on the chalice, the poet describes its reception by the priests of the temple:

> Et cil ki le Sepulcre gardent,
> Les reliques et le tresor,
> Ont pris le riche vaissel d'or.
> Ml't l'esgardent, cascuns s'en saigne
> Por la biauté et por l'ouvraigne
> Ki si est riche tot entor. (vv. 628–33)

> [*And the ones who watch over the Sepulcher, the relics, and the treasures took the precious vessel made of gold. They examine it closely, each one crosses himself in awe at the thoroughly fine beauty and workmanship.*]

For Linda Cooper, this otherwise traditional ceremony is quite "surprising" in light of the secular themes on the goblet. Loomis, likewise, saw a "startling" irony in the incongruities created by the scene (Cooper 158; Loomis 17–18). This is indeed a tantalizing scene because the impressive, potentially sublime, juxtapositions are neatly balanced by the psychology of protocol and diplomacy, by the opacity that conceals the priests' interior responses, should they differ from the outward display of reverence. Cooper, for example, suggests that the priests *must* be astounded by what *must* be deemed a quite unlikely offering. But the rich ironic potentials notwithstanding, the poet's syntax seems to run counter to any such conscious design. We are not told simply that the priests crossed themselves in reverence while receiving the offering but specifically "*por* la biauté et *por* l'ouvraigne / Ki si est riche tot entor." Of course the ambiguity surrounding the priests' appreciation of the details, of the exact content of the imagery, leaves the door open to further playful speculations (as Cooper also notes). The poet does say that they studied it closely however ("Ml't l'esgardent"). Perhaps the poet is suggesting that the priests merely turn a blind eye to the nature of the images themselves. Whatever the case, the author provides few clues to confirm any essentially ironic intent. What is clear is that Jean Renart takes up the thread of symbolism surrounding the love potion that we examined in Béroul. To the extent that the confrontation with Ogrin brings into play numerous notions of "*peché*," so too does the taking of the potion evoke ambivalent, even opposing, symbolic registers: that of the Apple and original sin, on the one hand, and that of the Eucharistic wine, on the other. I have suggested that it is this second association which prevails along with the lovers' union in the end. With the description of the coupe d'or in *L'Escoufle*, this *symbolic* association is transformed into a *literal* gesture:

> Li quens lor prie par amor,
> Ains k'il s'en aut a son ostel,
> Que, por Dieu, sor le maistre autel
> Soit pendus cil riches vaissaus
> *Et cil par cui li mons ert saus*
> *I soit et mis et honerés.*
> "Sire, ja mar me parlerés
> Font il, tot iert a vo devis.
> Hui en cest jor i sera mis,
> Ja n'i querrons atendre plus :
> N'en doit douter ne vos ne nus." (vv. 628–644; emphasis mine)

> [*Before he goes to his lodging, the count asks that as a sign of mutual love and respect and in honor of God, the precious vessel be placed on the main altar and that he who will save the world be both portrayed and honored there. "Sire, no need to speak more, everything will be as you desire. This very day it will be placed there; we have no wish to put it off longer—neither you nor anyone has cause for doubt regarding this."*]

Perhaps it is this perfectly logical and no less startling transformation itself which gives the passage its provocative power. In light of the emerging synthesis of theological and secular ideals resulting from new marriage doctrine, this blatant convergence of secular and religious elements—appropriately situated in Jerusalem, the emblem of a *religious state*—need not necessarily be seen as inherently incongruent and ironic. Indeed, the image which ends the description of the golden chalice is that which recalls the demise of the treacherous dwarf who sought to discredit and scorn Tristan and Iseut (vv. 610–17). It is Jean Renart who, like his later counterpart Guillaume de Machaut, defines the very essence of nobility not by rank as such but in terms of an idealized amorous sentimentality and amorous esthetic. The juxtaposition of sacred and profane elements in the scene of the golden chalice provides a neatly condensed description of a culture now based on two distinct yet parallel "orders." Each order has its own central preoccupations and icons, but their rites and symbols share common features as part of a larger system of sacred meaning and religious experience united by one Christian destiny.

There is perhaps an undeniable *ironie du sort* in the very course of Western thought, in the profound reversal that ultimately takes place with regard to theoretical doctrine on love and marriage. However, this deeper irony of historic change need not invalidate the very plausibility of such cultural shifts. The true amorous experience is for the secular individual tantamount to the true religious experience of the religious disciple just as the meaning of the amorous martyr within the public arena parallels that of the religious martyr or saint. The very syncretism encapsulated by the placement of the golden chalice at the holy altar and by its potential role as a receptacle for the symbolic Host is fulfilled in the larger parallels between hagiographic tradition on the one hand, and the amorous trials and tribulations of Guillaume and Aélis on the other. In fact, critics have recognized that *L'Escoufle* has many close parallels with another hagiographic narrative, that of St. Eustace:

> En plus d'être un "roman de couple", qui—à la façon de *Cligés*—exploite la légende de Tristan, *L'Escoufle* est en même temps un "roman de jumeaux", qui comprend les éléments particuliers de miracle, de souffrances et de triomphe par lesquels il participe à la catégorie de l'*exemplum* du martyr. Et la légende de Tristan et la convention des jumeaux sont toutes les deux des transpositions séculières du paradigme hagiographique, lui-même une christianisation du héros martyr, cela ne nous étonne guère de trouver dans ce roman l'exploitation abondante et la référence obsessionnelle à la légende de Tristan unies avec la convention des jumeaux. Car, en troisième lieu, *L'Escoufle* est—comme *Guillaume d'Angleterre* et *Guillaume de Palerne*—une sécularisation de la légende de saint Eustache . . . . (Cooper 164)

> [*In addition to being a tale based on the couple motif, a tale which in the manner of* Cligés *exploits the Tristan legend,* L'Escoufle *is at the same time a tale based on the twins motif involving such specific story elements as the miracle, as well as the elements of suffering and triumph which include it in another category, that of the* exemplum *of the martyr. Moreover, both the Tristan legend and the twins motif are secular transpositions of the hagiographic paradigm, which is in itself a christianized variation of the hero-martyr motif; it is hardly surprising therefore to find in this text the frequent use of and the obsessive references to the Tristan legend combined with the twins motif. For,* L'Escoufle *is—as are* Guillaume d'Angleterre *and* Guillaume de Palerne—*a secularized version of the Saint Eustache legend. . . .*]

Since, however, this vision of love depends upon a supernatural perspective—a poetic, esthetic, and religious vision of life which is transported into a world of mundane wealth and urbane civility—there remains an inexhaustible potential for surprising juxtapositions of secular and profane elements. These juxtapositions may be intentional, the author's way of dramatizing the difference between old and new religious and cultural perspectives. Or these juxtapositions may be surprising only to modern readers who have defined twelfth-century attitudes according to excessively dogmatic or puritanical views. If restricted by a too reductive moral view of marriage and adultery, the critic inevitably defines the narrative perspective in terms of a narrowly defined clerical irony—where irony is censure.

In light of trends we have identified in twelfth-century marriage fictions, it is clear that the episode of the golden chalice in Jean Renart's *L'Escoufle* provides perhaps the most striking and succinct image we have yet encountered of the secular-religious "space" that emerges from the development of sacramental marriage doctrine. Obviously, the emergence of this new cultural space explains the close relationship between religion, *eros*, and *translatio imperii* in Old French marriage fictions, while also providing for perspectives which are neither exclusively allegorical nor exclusively ironic with regard to the juxtaposition of religious and secular "attitudes," icons, motifs and metaphors.

PARIS AND TRISTAN: NEW HEROIC PARADIGMS

> Et s'en amor a un mesfait
> Ces coses font ver Diu bon plait,
> Qu'il aime honor et cortoisie
> Et fine Larguece est s'amie
> *Eracle* (vv. 3715–18)

Along with the development of twelfth-century marriage doctrine and the revival of archaic abduction motifs we find, not surprisingly, the emergence of an alternate feudal heroic archetype. A number of Old French marriage fictions exploit, as literary-historical reference points, both the figure of Paris (initially rehabilitated in both the *Roman d'Enéas* and the *Roman de Troie*) and the figures of Tristan and Iseut.[38] Prominent examples occur in *Partonopeus de Blois*, *Eracle*, *L'Escoufle*, and *Guillaume de Dole*.

We will recall that Partonopeus's virtues as a knight are accounted for and emblematized by his Trojan descent.[39] In one passage, as Mélior addresses her champion Partonopeus, she makes reference to his Trojan ascendancy as she describes his physical beauty and his exceptional moral, as opposed to material, *honor*:

> Qar toute beautés vos abonde,
> Et tant avés gentelise,

Jà ne lairai ne vos eslise;
Car vos estes del sanc Hector,
Qui ainc n'ama argent ne or,
Ne rien fors seul cevalerie. (vv. 1498-1503).

[*For you abound in all forms of beauty and show such kind manners that
never would I consider other than to elect you, for you are of the blood
of Hector who in his time never loved silver or gold or anything other than
chivalry itself.*]

The Trojan association is reinforced again at the end of the *A* manuscript,
at the very moment Partonopeus emerges as the victor in the contest for
Mélior (Simons and Ely 12-13). It is King Lohier who identifies the victo-
rious champion:

Mes pères moult cier vos avoit,
Perdus fustes en son endroit;
Chacier estoit alés li rois
Ès grans forès en Ardenois.
Vos fustes fils de sa seror,
De le haute geste francor
Et del linage as Troiens,
Don't on parlera mais tostens. (vv. 9265–72)

[*My father held you as one very dear to him; he considered you as lost. He
had gone to hunt that day in the great Ardennes forest. You were the son
of his sister, descended from the high Frankish nobility and of the lineage
of Troy, about which people will talk forever more.*]

While this reference and the poet's somewhat original evocation of the Troy
story may serve to exploit a popular Plantagenet genealogical myth (here
in favor of the Court of Blois) (Simons and Ely 12–13), it is also apparent
that the Trojan topos serves to evoke and define a heroic typology that
associates imperial conquest and dominion with an individual ethic based
on beauty, fidelity, and amorous courtship.

Gautier d'Arras also evokes Paris's famous abduction as a subtext for his
tale *Eracle*. In a thorough source analysis, Fourrier demonstrates that
Eracle contains elements from a popular tale known as "The Three Gifts"
and from a pseudo-historical legend recounting the marital fortunes of the
Greek emperor Theodosius the Younger (Fourrier 216ff.). The adulterous
hero in the latter tale is based on the figure of Paulin—the adulterous hero
in the Greek legend of the Empress Athenaïs. By virtue of an incident, like
that which will recur in the thirteenth-century *La mort le roi Artu*, an apple
received as a gift leads to a false accusation of adultery between Paulin and
Athenaïs. The emperor has Paulin killed and repudiates Athenaïs. In
Gautier's tale, however, the figure Paulin is now named Paridès, and
Fourrier attributes the modification directly to Gautier. Through the evocation of the Greek hero Paris, the cherries which replace the apple are
reconnected with the motif of the Phrygian Apple of Discord (Fourrier

223–4). The evocation of the Judgment of Paris brings to mind the ensuing abduction of Helen and the fall of Troy.

It is also noteworthy that Gautier seems to provide for a striking omission by departing from the culminating episode in the Tale of the Three Gifts. In the known version, the sage demonstrates his ability not only to judge women, but to judge men as well—by attesting to the illegitimacy of the king figure himself. In Gautier this general motif is adapted to the outlines of the Athenaïs legend when a humble orphan woman is chosen as a fitting bride for the emperor (Fourrier 216–21). The revelation of the emperor's true origin has, however, no immediate corollary; yet one does emerge in Gautier's larger reworking of the historical legend. The latent illegitimacy of the king figure is mirrored in Gautier's work by the usurpation of the emperor's place as sovereignty figure by the young suitor Paridès (counterpart of the young prince Paulin in the Athenaïs legend).[40] The textual references to the tale of Athenaïs evoke the seminal tale of Paris and Helen in another way as well, within a larger historical and iconographic context. For, as Fourrier points out, Constantinople was itself alive to the memories of Paris, Venus, and the Phrygian Apple—all memorialized in statues (Fourrier 243).[41] Likewise, the Empress Athenaïs (Eudoxie) was vividly remembered for her generous gifts to Jerusalem and referred to as "the new Helen" (Fourrier 244).

A less substantial but no less interesting web of intertextual references to the figures of both Paris and Tristan can be found in the tale *L'Escoufle*. We commented earlier on the *Tristan* episodes engraved on the golden chalice that Count Richard (the hero's father) offers at the altar of the Holy Temple in Jerusalem. In relation to the story's central amorous hero (Guillaume), the chalice episode could be said to anticipate the iconographic importance of the figures of Tristan and Iseut in this tale of erotic destiny and amorous devotion. At one juncture in his frustrating search for Aélis, Guillaume makes a poignant appeal to Saint Gilles who then answers his prayers and guides the next stage of his quest (vv. 6466–6503). In a similar manner, Tristan and Iseut serve as literary icons or "amorous saints" representing a mysterious and providential aspect of secular experience (e.g., vv. 3450–65; 7820–3). In *L'Escoufle*, therefore, the hero's quest is modeled on amorous and hagiographic motifs, and the figures of Tristan and Iseut become emblems of secular martyrdom.

Jean Renart also makes allusions to the legacy of Troy in his depiction of the lovers' eventual marital triumph. Shortly before the long-deferred reunion of Guillaume and Aélis, the narrator conveys the joy of the ensuing event by evoking the abduction of Helen (". . . puis qu'Elene vint par nage / De Grece en la terre Paris, / N'ot tant de joie ne tant ris" vv. 7674–76). When Guillaume's identity is revealed to the court society where Aélis has been retained, the Count of St. Gilles announces that Guillaume's father (Count Richard) was his own cousin's son and that he will see to it that Guillaume regains his rightful domain ("Je vos metrai en vo demaine / Tot Rueem et Moustierviler" vv. 7750–1). During the subsequent festivities in celebration of Guillaume's new wealth and title, the narrator once again alludes to the days of Troy:

> En cele quinzaine u il ere
> Fu Guilliaumes fais chevaliers;
> Li quens l'i fist ml't volentiers
> Et bien .xxx. autre[s] pour s'amour.
> La joie, la feste et l'ounor
> Ne sai je pour coi j'acontasse:
> Puis que Troie la grans fu arse
> N'ot il a .j. chevalier faire
> Tant de dames de haut afaire
> Ne tante de pucele de pris. (vv. 7902–11)

> [*In that fifteenth year of his William was made a knight quite willingly by the count, along with some thirty other young men. I do not know to what I might compare the joy, the celebration, and the honors. Since the burning of the great city of Troy, never have there been so many ladies of high distinction and young girls of such merit present for the making of a knight.*]

It is remarkable that a ceremony emphasizing male bonds and male initiation—based on the values of Dumézil's first and second functions—is here measured according to the presence of refined ladies and maidens. At the same time, the poet inserts an elliptic allusion to the past which reminds us that the amorous courtship between Paris and Helen (figured in the medieval tale as a conquest based on mutual and providential love, not as rape) moved both armies and empires.

Of course, reference to the burning of Troy also augurs the transfer of imperial power to Rome in the present narrative. The allusion anticipates Guillaume's own impending triumph, explicitly foreshadowed during the same episode:

> Font li chevalier : "Ore es che
> Bons commencemens de jovene home.
> Encore iert il sire de Rome,
> Si Dieu plaist, et sa feme vit." (vv. 7924–27)

> [*The knights said, "Now has come to pass a young man's auspicious beginning. Yet will he become ruler of Rome—if it please God and his wife lives."*]

It is interesting that the poet includes Guillaume's wife in his reference to the providential conditions governing the ensuing political triumph. The significance of her role remains vague, but so are the mysteries of God's will evoked in the same line.

At the end of the narrative, therefore, the parallel between Guillaume's fortune and the heroic age of antiquity becomes almost literal. Of course a certain distance is also established: while the historic parallels define the archetypal dimensions of Guillaume's amorous quest, the hagiographic parallels add a new dimension to the heroic paradigm. What is quite clear, however, is that the heroic model to which the hagiographic elements adhere departs from the precedents of classical epic—from the prestige accorded the first and second (king and warrior) functions described by Dumézilian tripartition. References to the past remain rather narrowly centered on events relating to Paris's abduction of Helen. The nature of Paris's persona as well as the privileged role given to Venus in the tale of Paris's judgment identify Paris as a figure associated with third category values and functions, with esthetic and amorous (as opposed to martial) values. At the same time, the abduction of Helen is a mythical topos which most dramatically reflects the inherent relation between amorous and political sovereignty. In this respect the Greek tale of bridal abduction allows for a ready assimilation with Celtic matter. For the figure of the sovereignty goddess in larger Indo-European tradition clearly informs the particular Greek myth of Paris and Helen just as it plays a quite dominant role in the surviving body of Celtic myths and legends. The fact that the author of *L'Escoufle* has selected Paris and Tristan as the central intertextual reference points clearly underlines just how consciously medieval thinkers turned to these traditions as relevant *exempla* in relation to contemporary speculations with regard to marriage doctrine and the role of marriage in relation to temporal values, *translatio imperii*, and new heroic archetypes.

However, the close typological correspondence between Paris, Tristan, and the fate of Guillaume in *L'Escoufle* (all revealing similar relations among love, sovereignty, and *translatio imperii*) does not so fully obtain in the case of Renart's allusion to the Troy story in his subsequent work *Le Roman de la Rose ou de Guillaume de Dole*. In this tale the main amorous protagonist, the German emperor Conrad, already enjoys an uncontested and prosperous imperial sovereignty. Nonetheless, at the very moment of his conjugal triumph, the author goes out of his way to evoke memories of the Trojan myth. We recall that at the beginning of *L'Escoufle* the narrator describes a golden chalice inscribed with images from the legend of *Tristan et Iseut*. In *Guillaume de Dole*, a similar device is used. Images evoking Paris's abduction of Helen are embroidered on the bride's (Liénor's) gown:

> D'un drap que une fée ouvra
> Fu vestue l'empereriz;
> Il n'iert ne tissuz ne tresliz,
> Ainçois l'ot tot fet o agulle
> Jadis, une roïne en Puille,
> En ses chambres, por son deduit.
> El i mist bien .vij. anz ou .viij.
> Ainz que l'oevre fust afinee.
> Einsi com Helaine fu nee
> I estoit l'istoire portrete;

Ele meïsme i fu retrete,
Et Paris, et ses frere Hectors,
Et Prians, li rois, et Mennors,
Li bons rois qui toz les biens fist.
Et si com Paris la ravist
I sont d'or fetes les ymages,
Et si come li granz barnages
Des Grieus la vint requerre après.
Si i fu aussi Achillès
Q'ocist Hector, don't granz diels fu;
Et si com cil mistrent le fu
En la cité et el donjon,
Q'en avoit repost a larron
El cheval de fust, et tapis.
En ce qu'il jut sor les tapis,
Desroulee fu la navie
Des Grieus : pieça or n'est en vie
Hom que si biau drap seüst fere. (vv. 5324–51)

[*The empress was draped in a dress stitched by a fairy. It was neither woven nor braided; rather, in days gone by a queen in Puille made it entirely with a needle to distract herself as she sat in her chambers. She spent a good seven or eight years working on it before the final touches were done. One could see depicted on it the events leading up to Helen's birth as well as a portrait of Helen herself, along with Paris and his brother Hector, Priam the king, and Menor [Memnon?], the good king who did so many fine deeds. And the story of how Paris abducted Helen is there in images embroidered in gold, and also there the story of how the great Greek nobles came to demand her return. Achilles was depicted there as well, who killed Hector and caused great mourning. And the manner in which the Greeks who had hidden themselves treasonously in the wooden horse put to fire the city and tower was also depicted. And spread across the bottom folds of the dress was the Greek fleet. It has been a long time since there lived anyone able to make such a beautiful dress.*]

Symbolically, the wedding gown reflects the triumph of the values exhibited at Conrad's court and adds a further esthetic and cultural richness to Conrad's and Liénor's marital success. Literally, however, there remains no clear analogy between the emperor's amorous conquest and the epic legend of Troy. In *Guillaume de Dole*, therefore, we witness a shift from a truly epic accounting of amorous courtship, an accounting based on enduring epic aspirations, to an ideological framework in which time remains static and "reality" becomes primarily a function of esthetic and imaginative perception. In other words, Jean's work introduces the medieval audience to the world of the timeless idyll, like that encapsulated in the allegorical garden of Guillaume de Lorris's *Roman de la Rose*. In this sense, Renart's text represents the culmination of the ascendancy of the medieval Paris figure. While Paris's judgment motivates an epic battle and transfer of empire, we should recall that the figure of Paris and the legend of Paris's judgment suggest a different line of Western ideology, one which reflects primarily neither sovereign dominion nor military prowess but pastoral tranquillity and physical beauty. Conrad's court and his leisurely forest picnic (vv. 138–558) represent a return to the static and pastoral setting of Paris's judgment, away from monumental historical progress.

Of course this view of *Guillaume de Dole* is entirely consonant with the treatment of love and esthetics throughout this particular tale. For what Jean Renart seems bent on conveying is the extent to which amorous sentimentality represents an imaginative construct closely tied to esthetic cultivation and refinement. We recall that the Emperor Conrad falls in love with Liénor, not at first sight but rather at first mention. Not only does the expression of sentimentality remain tied to lyricism throughout the narrative, but the very initiation of Conrad's amorous infatuation is linked to the idle storytelling of his court *jongleur*. Accordingly, the reign of the German Emperor Conrad is marked by sumptuous living and a refined esthetic sensibility. In this respect, Jean's vision of love, courtship, and marriage distances itself from the kind of epic and hagiographic exploits and conflicts depicted in the earlier narratives we have examined.

NOTES

1. As for Chrétien's satirical treatment of amorous excess, it would seem to suggest a discretionary attitude toward secular love and courtship. We will discuss this further in our remarks on *Cligès* below. In works by other authors—works such as *Eracle* and *Flamenca* for example—we see an explicit distinction as well between "folie" and "Amor," between coercive passion and jealousy on the one hand, and a mutual affection and natural compatibility on the other.

2. *Guillaume de Dole* is perhaps the only exception in that no clear customary or corporate interest, but rather a kind of private malice, accounts for the obstacles that would derail the emperor's plan to marry Guillaume's sister. I will discuss this work below, however, in other contexts.

3. Chapter Two contains a close study of the *Enéas*.

4. The opening of the tale of "Guigemar" by Marie de France provides a noteworthy example of contemporary attitudes toward Ovid. The author rather explicitly alludes to and critiques Ovidian psychology, neither accepting nor rejecting it wholesale.

5. In Chapter Three we saw that the hierarchical model of ethical judgment had an explicit foundation in the distinction to be made between customary laws, on the one hand, and the notion of natural or universal laws, on the other. Of course, as was clear in the *Roman d'Enéas*, the ethical ambiguities inherent in this theory of standards can be exploited according to a convient political or ideological *parti pris*. In terms of narrative structure, we saw in our study of the *Enéas* (Chapter Two) that the topos of the *vox populi* is adopted from Virgil as a means of representing and contrasting ethical perspectives. It reappeared in Benoit's reflections on Crisede's departure from the Trojan citadel (Chapter Three). Below I will discuss innovative uses of the topos in subsequent Old French marriage fictions.

6. See Chapter Three.

7. Of course, the name is not insignificant. According to Fourrier's analysis of Gautier's probable sources, the allusion to the Trojan hero is to be credited to Gautier himself (Fourrier 222–4). I will discuss this point further below.

8. I will discuss both works in greater detail below.

9. Flamenca is likewise eclipsed from the preparations for the marriage which are also carried out according to strict conventions (e.g., reference made to the initial deliberations between Count Guy and Archimbaut's emissaries and the subsequent preparations, vv. 70ff).

10. Indeed, Partonopeus's own initial "rape" of Mélior not only underlines the fallacious dichotomy separating the two worlds, but "marks" Partonopeus as an initiate still steeped in patriarichal mores. Partonopeus's contact with Mélior ultimately results in a deeper and more sacred form of amorous experience leading to a consensual marriage and an affectionate marital bond.

11. The reference to Hector and the Trojan line is of course not insignificant in light of our larger analyses. I will comment on this and other allusions to Paris, Helen, and Troy later in this chapter.

12. For the controversy over the work's original unity and composition, see (in addition to Louis) J. Bédier, *Les légendes épiques*, 3rd ed. (Paris: Champion, 1926) 1: 3–95; F. Lot, "Etudes sur les légendes épiques françaises, II: *Girart de Roussillon.*" *Romania* 52 (1926): 257–90; P. Le Gentil, "*Girart de Roussillon*, sens et structure du poème." *Romania* 78 (1957): 328–89; 463–510.

13. The first application of the terms "epilogue" and "prologue" actually goes back as far as Paul Meyer's introduction to his translation of the poem: *Girart de Roussillon* (1884; rpt. 1970). I include a brief summary of the plot in my remarks on *Girart de Roussillon* in chapter one.

14. Carles est erberjaz sobre Saoune;
Tiebert de Vaubeton pres per la gone,
Isenbert e Brochart, cui enrazone:
"Grant aver a Girarz e terre bone.
Des le Rin tec s'onor trosque a Baioune,
E devise Espaigne per Barçelone,
E li rendent treüt cil d'Arragone.
A! com es fols lo reis qui tau fiu done!
E qui aleu m'o quert, lai m'arazone.
Lo reiame desfait e despersone;
Eu non(c) ai plus de lui fors la corone.
Mais eu li cuit mermar tro a Garone." (vv. 556–67)

15. In Chapter One I discuss briefly the limited influence of troubadour models on the amorous intrigue in *Girart de Roussillon.*

16. An overview of the amorous intrigues in the Oxford version of the poem is included in my discussion of *Girart de Roussillon* in chapter one.

17. This claim is clearly demonstrated by de Combarieu's point by point comparison and contrast of the marriage plot in *Girart de Roussillon* (Oxford version) with the comparable plot circumstances presented in *Garin le Lorrain* (Combarieu du Gres *Elissent* 25).

18. Cf. our analysis in Chapter One.

19. For an example of one scholarly critique of Duby's reductive opposition see Colish (2: 658).

20. Louis, by contrast, sees the Pope depicted in a much more positive light (2: 377).

21. In Chapter Three, for example, I pointed out the extent to which theological definitions of marital consent drew upon the interests of contemporary ethical philosophy, the concept of *intentionality* in particular. The intentional and subjective disposition of the individual is contrasted with external words and action. See also Hunt's article on Béroul's *Tristan et Iseut* and our discussion below.

22. In *Girart de Roussillon*: 1) the importance of the sovereignty rivalry, already inherent in the conventional epic topos of the rebellious baron, is singularly underscored in the "prologue" portion of the poem when, in a line we cited earlier, the king expresses his concern over Girart's growing power: "Eu non(c) ai plus de lui [Girart] fors la corone" l. 566. 2) Girart's reputation as a war hero precedes his arrival in Constantinople where he is to be betrothed. 3) Like numerous sovereignty bride / mother figures (as clearly reflected in extant Irish sagas for example), Elissent plays a pivotal role in reconciling Girart and King Charles, and the restoration of land to Girart's control depends upon her authority and upon Girart's twin marriage alliances. 4) We have analyzed in detail the two marital bonds linking Girart to both Berthe and Elissent. For a recent, succinct and comprehensive overview of the main features of the Indo-European sovereignty myth with a detailed survey of its Celtic manifestations, see Kim McCone, *Pagan Past and Christian Present in Early Irish Literature* (Naas, Co. Kildare, Ireland: An Sagart, 1990; rpt. 1991), chapter 5.

23. We should recall that the implicit Christian "reading" of the *Tristan* exile motif is provided by analogy with hagiographic texts such as the *Vie de St. Eustache* and especially the *Vie de St. Giles* (cf. Lewes and our discussion of *Tristan et Iseut* below).

24. Hence, I agree with Deborah Nelson's reading of Chrétien's irony: "In fact, within the framework of the plot, Fénice's deeds receive even more attention than those of Iseut by the world at large because of the spectacular ruse of the false death" (83). Shirt, for example, has argued that, technically, Fénice is absolved of any wrongdoing, to the extent that she avoids any objective transgression of doctrinal prescriptions as such. Of course, as Nelson points out, "[Fénice's] insistence on giving 'corps et coeur' to the same man is based totally on her desire to maintain a virtuous public image and not on a moral aversion to adultery . . . The technical difference between the two women—that Fénice's and Alis's marriage was never consummated—made absolutely no difference to posterity" (83). I will discuss Fénice's attitude toward public opinion in more detail below.

25. Chrétien de Troyes, *Cligés*, edited Stewart Gregory and Claude Luttrell (Cambridge, England: D. S. Brewer, 1993), p. 273.

26. Motifs in and allusions to the Old French *Tristan et Iseut* provide an intertext for numerous Old French narratives including several works discussed here (*Cligès*, *Girart de Roussillon*, *L'Escoufle*, *Flamenca*) (cf. Kay).

27. As the editor Philippe Walter points out in his note to Béroul's text: "Ce dialogue est placé sous le signe du malentendu. Les mots *foi* et *péché* n'ont pas le même sens pour Ogrin et Tristan. Leurs connotations chrétiennes ne sont nullement perçues par Tristan. *Péché* en ancien français signifie aussi "erreur", sans référence à une transgression de la loi divine."

28. I will soon suggest, however, that Béroul's perspective also evokes an opposing analogy, that between the potion and the Eucharistic wine.

29. It is this broader conclusion which allows us to reject a Robertsonian approach, based on an ironic inversion of Christian morality, while not asserting the existence of a purely "secular" tradition or genre. It is also worth recalling here a point made in Chapter Three, that the emergence of a revised marital canon coincides historically with reforms specific to monastic morality—reforms which emphasize a strict doctrine of celibacy.

30. Since Béroul's earlier description describes the lovers as being immune to pain (*dolor*), the *paine* referred to here would seem to denote material deprivation more than physical and emotional distress; cf. our discussion of Thomas's version which follows.

31. Peter F. Dembowski provides a larger overview of the influence of vernacular hagiography on a number of twelfth- and thirteenth-century secular narratives. This influence results, according to Dembowski, in the "'hagiographization' of the hero in the literary work" (p. 119). See "Literary Problems of Hagiography in Old French," *Medievalia et Humanistica*, n.s. 7 (1976): 117–130.

32. Translated by Daniel Lacroix (see Works Cited under *Tristan et Iseut*).

33. I cite this passage in Chapter One. An identical lament appears in the text of *Eracle* at the moment when Paridès's crime is uncovered:

Li varlés est molt angosseus
et angossant va li espeus
et molt angossant vait l'espeuse,
gens ne fu mais si angosseuse.
Li varlés crient, li sire plaint
et li dame palit et taint ;
li uns se plaint molt durement,
li doi ont paor de torment,
que gerredons lor soit rendus,
qu'ele soit arse et il pendus. (vv. 4887–96)

34. Unfortunately, this passage is not found in the 1834 Crapelet edition which is based on a manuscript with a revised ending (Fourrier's ms A.). My source for the citation is Fourrier's chapter on *Partonopeus de Blois*. Fourrier has extracted the passage from a fourteenth-century manuscipt at the *Bibliothèque de Tours*. This is Fourrier's ms *T*.: "Le seul allant jusqu'au bout du récit. En revanche ne commence qu'au v. 105. Déjà signalé en 1834 par Raynouard dans le *Journal des Savants*, p. 731, n. 1, mais sans qu'en ait été reconnue toute l'importance. Souvent peu sûr dans ses leçons, mais très précieux" (315, note 1).

35. I say foils and not truly antagonists because both are depicted as sympathetic, less fortunate companions of Partonopeus—companions in the sense that they, like

the main hero and the narrative persona, are all (to use Dante's expression) captive souls and gentle lovers on Love's road.

36. For manuscript and source, see note 33 above.

37. On *Cligès*, see Peter Haidu, *Aesthetic Distance in Chrétien de Troyes: Irony and Comedy in Cligès and Perceval* (Geneva: Droz, 1968). With regard to *L'Escoufle*, I will focus my remarks on the significance of the "coupe de Tristan." For an excellent study of this motif and ironic readings of *L'Escoufle*, see Linda Cooper, "L'Ironie iconographique de la coupe de Tristan dans *l'Escoufle*," *Romania* 104 (1983), pp. 157–176.

38. In *L'Escoufle* we will see a regular alternation between the two figures as textual reference points.

39. The fictitious genealogy is the subject of the prologue.

40. As for the status of the empress as a sovereignty-bride figure, this is evidenced in lore by references to her as *Augusta* (cf. Fourrier, pp. 220–1).In the text of *Eracle*, see Paridès's speech with regard to the emperor's inflated status, a status based purely on title and not heroic merit (vv. 4636ff.).

41. It should be mentioned that the golden apple has associations with sovereignty (e.g. *Chanson de Roland*) and appears in the form of a golden globe in the medieval iconography of Constantinople (see Fourrier, p. 243). In fact, Fourrier points out that Gautier, in his references to the relevant statue, once again elects to omit any mention of the golden apple. Is it possible Gautier sought to dissociate the adulterous hero from the moral censure attached to the apple motif in biblical tradition?

Western Tradition and the Old French *Roman*

The highly esthetic expression of sentimentality in *Guillaume de Dole* gives one the impression that the amorous adventure is merely one model among others giving form to a deeper ideological shift. In this sense *Guillaume de Dole* has a singular place among the works we have cited. Although it reflects an already established shift in emerging views of secular ideology, it gives sentimentality a truly prominent role by rejecting competing generic models (e.g., hagiography and epic) and highlighting contemporary lyrical models. However, it goes so far in its esthetic vision of love that it also suggests that love may be, like poetic imagery, a purely esthetic and *cultivated* form, as opposed to an ontological reality. This represents a real or potential departure from the underlying mythical and theological views of *eros* that define twelfth-century assumptions concerning the sacramental value of marriage. Despite this departure, however, *Guillaume de Dole* reflects an acute insight into larger ideological tensions and shifts, between competing Western archetypes on the one hand and competing models of sovereignty on the other—one model based on objectively defined forms of rank and power, and another based predominantly on subjective and esthetic imagination.

It is perhaps not surprising that the development of a theological theory pertaining exclusively to the secular order would not only draw on secular traditions and precedents but ultimately result in something akin to a new and largely autonomous secular ethic. Clearly, however, the emergence of a new Christian marriage doctrine endowed the dominant motifs and themes of secular lyricism with greatly expanded ideological dimensions. Finally, the relation between love, imagination, and fiction reflects a culminating point in the development of an ideological esthetic already expressed by Ovid and culminating in the metaphorical forms of elaboration which give the thirteenth-century Provençal text *Flamenca* its highly stylized quality.

Our examination of medieval marriage fictions can in fact help us to see the development of the medieval *roman* within a larger context of literary history. The very tension between natural and esthetic definitions of *eros*, between epic and pastoral aspects of the Paris myth—adumbrated in the progress from the *Romans d'antiquité* to *Guillaume de Dole* and *Flamenca*—has roots in Ovid and reveals itself in the mixture of epic, fabliau, and hagiographic elements that characterize the Old French portrayals of Tristan and Iseut. In the Old French *Tristan*, as in Ovid, the amorous itinerary reflects an all-encompassing and civilizing sublimation of more rudimentary secular struggles and impulses. The theatrical nature of this sublimation is a central aspect of Ovid's ideology, and the importation of *fabliau* elements in the *Tristan* narratives echo the Ovidian delight in dissimulation and play as positive features of aristocratic refinement.

In Book I of the *Ars amatoria*, Ovid defines love paradoxically as a natural instinct and as a cultivated form of metaphoric drama. Accordingly, love is not only a distinct skill or "science" but a virtue whose cultivation becomes the centerpiece of cultural progress and refinement. In both Ovid and the Old French *Tristan* tradition, we see a common emphasis on the natural or transcendent authenticity of erotic desire. The point merits our attention for at least two reasons. First, it provides for a complementary convergence of Ovidian love (and other forms of cultivated lyricism such as twelfth-century *fin'amor*) with the emerging view that the bond of marital affection is an extension of a universal impulse governed by natural law to the same extent that (under the new dispensation) this same affection derives from the divine grace inherent to the larger operations of marriage as a sacrament. Instances of pure deception aside, various manifestations of *eros* in various contexts all conceal an underlying ontological impulse. Secondly, therefore, the assumptions that underlie the treatment of love in medieval texts may essentially depart from modern deconstructionist views which tend to privilege the artificiality of linguistic and ideological conventions at the expense of univeral laws and instincts. In Stendhal's *De l'amour*, for example, amorous fascination is clearly depicted as a psychological fixation (*cristallisation*) whose catalyst is an immediate and primal erotic impulse. In our century, by contrast, literary critics have abandoned the concept of an authentic amorous impulse and defined it as a purely mediated experience—as an act of subjective fantasy and desire (René Girard), as an extension of verbal and narrative convention (Roland Barthes), or as the manifestation of competitive "homosocial" interaction between men (Eve Sedgwick).[1]

In his examination of love in *Girart de Roussillon*, for example, Simon Gaunt has argued that the marital conflict involving Girart, Charles, Elissent, and Berthe is entirely motivated by the political rivalry between the King of France and his vassal Girart. Accordingly, Gaunt contends that the imperial brides are not the objects of sincere desire and have a merely passive role as pawns in a "homosocial" contest of will and rank between a king and his most powerful baron. In his search for a suitable model to account for this kind of triangular mediation in which an amorous rivalry is motivated by a notion of "subjective rank," Gaunt alludes to the influence of troubadour lyric and *fin'amor*. While Gaunt's observations are consistent with our own initial analysis of the poem's preoccupation with the struggle for feudal honor, we have seen that the marriage fiction elaborated in the Oxford version does not really reflect the conventions of *fin'amor* as such. Furthermore, the "homosocial" theory of gender relations evoked by Gaunt would fail to account for the sincere expressions of mutual affection expressed by Elissent, Berthe, and Girart as well as the authentic sentiments which ratify the clandestine marriage between Elissent and Girart (not to mention the intimate bond forged in isolation by Fouque and Aupais). Indeed, we have seen that only in the limited case of the king's abuse of power does an external feudal rivalry govern the marital conflict. The poem's essential vision proposes a directly antithetical view in which the very authenticity of the amorous sentiment insures the integrity of the subjective forms of honor that safeguard the most basic feudal bonds.

Despite the underlying insistence on the authenticity of *eros*, however, we see that from the points of convergence between the lyrical Ovid and the Old French *Tristan* emerges a larger tradition which envisions a shift away from instinct and nature to a more sophisticated form of affective engagement in which courtship and love provide for the cultivation of a superior secular ethic—one anchored, nonetheless, in natural law and reflecting a divine dispensation, pagan in Ovid and Christian later. The reaffirmation of *eros*, therefore, also involves a new emphasis on esthetic refinement and subjective experience (over literal and external reality). This new ideological vision exploits traditional values as well, however, and Ovid demonstrates the historical and ontological continuity of the new amorous ethic in what is otherwise a purely humorous digression on the contemporary Roman theater as a good hang-out for randy bachelors. Excellent as it may be as a site of social dalliance, the Roman theater also has inscribed within it the memory of a more primitive and literal act of sexual and political conquest that is virtually coterminous with the founding of Rome itself:

> Primus sollicitos fecisti, Romule, ludos,
> > Cum iuvit viduos rapta Sabina viros.
> Tunc neque marmoreo pendebant vela theatro,
> > Nec fuerant liquido pulpita rubra croco;
> Illic quas tulerant nemorosa Palatia, frondes
> > Simpliciter positae, scena sine arte fuit;
> In gradibus sedit populus de caespite factis,
> > Qualibet hirsutas fronde tegente comas.
> Respiciunt, oculisque notant sibi quisque puellam
> > Quam velit, et tacito pectore multa movent.
> Dumque, rudem praebente modum tibicine Tusco,

Ludius aequatam ter pede pulsat humum,
In medio plausu (plausus tunc arte carebant)
    Rex populo praedae signa petita dedit.
Protinus exiliunt, animum clamore fatentes,
    Virginibus cupidas iniciuntque manus. . . .
Romule, militibus scisti dare commoda solus.
    Haec mihi si dederis commoda, miles ero.
Scilicet ex illo sollemnia more theatra
    Nunc quoque formosis insidiosa manent. (vv. 101–16; 131–4)

[*You, Romulus, were the first to disrupt the games when the rape of the
Sabine women served to console the widowed men. Not then did any cur-
tain hang over the marble stage, nor were the platforms red with flowing
Crocus. There, crudely dispersed, were leafy wreaths which the woody
Palatine had borne; the stage was simple, unadorned. The people sat on
terraced benches made of turf, an occasional leaf shading disheveled hair.
They look around, each one noting with his eyes the woman he desires,
arousing much emotion in their breasts. Then, to the crude melody of the
Tuscan piper, a player thrice stomps the level ground, and amidst the
applause (the applause in those days lacked any sophistication) the king
gave to the crowd the sought after signal that the spoils were at hand.
Straightaway they rush from the stands, their shouts revealing their excite-
ment, and lustfully lay their hands upon the maidens . . . You alone,
Romulus, knew how to comfort your soldiers—I too will be a soldier if
you should offer me such comfort. And thus from these hallowed customs
our theaters today remain fraught with danger for beautiful women.*]

In a paradoxical way, therefore, Rome's defining ideology ostensibly
remains intact over time but with the evolution from rustic to urban values
it comes to be enacted dramatically, socially, and metaphorically. The
attributes of the "warrior" (Dumézil's second function) are now co-opted
by the lover and seducer (associated with esthetic beauty and sensual pleasure, Dumézil's third function). Hence, the amorous ethic deceptively integrates old values at the same time that it mocks and subverts the ideological *status quo*.

This subversion is reflected in other gestures as well. Various *loci* of
Roman dynastic and imperial prestige are consistently transformed, one by
one, into a particular *locus amoenus* for amorous conquest. The porticos
and colonnades that are memorials to leading Roman figures, for example,
are touted as excellent sites for courting Roman maidens (vv. 67ff.). As suggested by the theater trope, these transformations represent not only an
ideological vision but a real historical evolution. Ovid suggests that the
past exercise of real power and conquest which made Rome into a virtually unchallenged imperial power is now (re-)enacted as dramatic repetition.
Reference is made, for example, to Augustus's recent staging of the historical battle between Persian and Athenian vessels at Salamis. The implication, of course, is that contemporary imperial Rome has entered into a new
cultural age privileging new forms of skill and action. Here the staging of

a monumental historical battle serves to encapsulate the image of Rome as the center of history and imperial dominion and to make Rome into a cosmopolitan host of nations. However, Ovid swiftly transforms the conventional political and military notions of conquest into amorous ones. Recommended by the exceptional number of foreign women with exotic appeal among the flock of spectators, the event is remembered as an exemplary occasion for seductive pursuits, thus subverting, mocking, and at the same time appropriating the emperor's own attempt to reaffirm, in an age of peace, the ideology of military prowess and conquest to which he owes his prestige (vv. 171–176).

In other words, the experiential and subjective reality of love provides for a theatrical and metaphoric integration in present time of historical and cultural values gained over time through material conquest. The underlying opposition, of course, is between the wild and the cultivated which defines the very ascendancy of Rome from its crude Tuscan roots to a sophisticated urban and imperial empire through a process of agricultural and military conquest. For Ovid, the cultivation of urbane love encapsulates the same dynamics, involving the domination and cultivation of violent passions and impulses as well as metaphoric links between such fundamental social activities as hunting and warfare.

Although love is a secret, affective, and subjective activity, it is not necessarily marginal and anti-social. Ovid defines amorous delight as a national imperative and as the exploitation of a national treasure:

> Gargara quot segetes, quot habet Methymna racemos,
>   Aequore quot pisces, fronde teguntur aves,
> Quot caelum stellas, tot habet tua Roma puellas:
>   Mater in Aeneae constitit urbe sui. (vv. 57–60)
>
> [*As numerous as the crops upon Gargara and the grape clusters that grow in Methymna, equal to the fishes in the sea and the birds hidden in the trees, as many as the stars of heaven, that is how many young girls your Rome has: the mother of Aeneas still makes Rome her home.*]

Humorous and picturesque, the harvest and resource analogy also fits into a larger ideological vision which attempts to reveal and encourage a shift from literal forms of heroic questing to a metaphoric form of conquest adapted to a new age marked by imperial stability and urban luxury:

> Tu quoque, materiam longo qui quaeris amori,
>   Ante frequens quo sit disce puella loco.
> Non ego quaerentem vento dare vela iubebo,
>   Nec tibi, ut invenias, longa terenda via est.
> Andromedan Perseus nigris portarit ab Indis,
>   Raptaque sit Phrygio Graia puella viro,
> Tot tibi tamque dabit formosas Roma puellas,
>   'Haec habet' ut dicas 'quicquid in orbe fuit.' (vv. 49–56)

[*You too, who seek matter for a lasting love, learn first where it is the young girls congregate. I will not order you to put sail to the wind in your search. Nor, as you will see, is there any long road to be trammeled. While Perseus may have brought Andromeda from the land of a black race and the Grecian girl may have been abducted by her Phrygian suitor, no less will Rome itself provide you with beautiful young girls—so many and so beautiful that you will say: "Right here one finds all that the world has to offer!"*]

The reference to Paris and the rape of Helen reminds us of the way in which amorous conquest remains linked to other forms of conquest and domination in the Western paradigm of imperial and cultural progress. To the extent that *any* amorous conquest involves similar constituents and challenges, Ovid allows for the assimilation of fundamental imperial values to a subjective form of cultivated love. Hence, Ovidian love represents a cultivated social ideal which is nonetheless rooted in natural impulses for which it substitutes and from which it extends. The Ovidian tradition, therefore, tends to synchronize and integrate the values delineated by Dumézilian tripartition as it envisions a real sequential progress from martial violence to amorous dalliance, from the ascendancy of first and second category functions to the ascendancy of third category functions. Just as Paris effects a transfer of dominion through the swift seduction of Helen, so do love and lyricism have the power to subdue the natural ferocity of the warrior (vv. 7–12). Simply put, the transformation of value involves a corresponding shift in ideological paradigms and in heroic models. As in the *Tristan*, therefore, the amorous "rule" incorporates the divine, violent, and instinctive aspects of *eros*, while effecting a transformation of value away from objective forms of social conflict and hierarchy and toward affective, subjective, and metaphoric forms of feudal harmony and engagement. Of course, the proposed transformation ultimately remains an idyll, and intransigent social attitudes and proscriptions put the amorous subject at risk—a fact to which the fortunes of the figures of Tristan and Iseut and, ironically, those of Ovid himself testify.

The emergence of a more distinct secular state and the rise of secular lyricism in the twelfth century involve a similar form of ideological shift. We have already seen how the rehabilitation of Paris and the re-evaluation of the amorous hero in twelfth-century retellings of the *Aeneid* and Trojan war legend served as part of a larger dynastic and political myth at the service of a sophisticated Norman-Angevine court. The revival of Celtic legend and the near-ubiquitous role of the goddess figure as a sovereignty bride allowed, of course, for the convenient intersection of amorous and political themes. Historically, however, we have seen that emerging marriage doctrine provided the basis for an even more profound shift in assumptions underlying secular values, as love and marriage (i.e., marriage properly defined and sanctioned by authentic forms of individual affection and consent) came to be defined as the primary entities governing the very destiny of the secular subject and of secular society and history as a whole. Accordingly, a secular ethic emerges with its own discursive categories distinct from orthodox patristic morality but not incompatible with emerging ecclesiastical doctrine. This larger transformation seems to account, at least in part, not only for the prominence of love and marriage plots in Old French secular narratives, but also for what Kay refers to as the "hybrid" nature of the Old French *roman*. The emergence of a distinct secular sacrament and concomitant secular "rule" centered on marital fortunes would seem to have prompted not only the development of an inclusive genre which draws inspiration from epic, lyric, and *fabliau* traditions but which also involves the regular integration of contemporary hagiographic motifs and models. Ultimately, it is the reform of aristocratic marriage practice that unites these diverse models into a generic field of narrative fictions that portray love and consensual marriage as the cornerstones of secular dominion, social harmony, and cultural prestige.

NOTES

1. René Girard, *Mensonge romantique et vérité romanesque* (Paris: Grasset, 1961); Roland Barthes, *Fragments d'un discours amoureux* (Paris: Ed. du Seuil, 1977); Eve Kosofsky Sedgwick, *Between Men: English Literature and Male Homosocial Desire* (New York: Columbia U P, 1985).

Bibliography

Adler, Alfred. "Eneas and Lavine: *Puer et Puella Senes.*" *Romanische Forschungen,* 71 (1959), 73–91.

_____ . "*Militia et Amor* in the *Roman de Troie.*" *Romanische Forschungen* 72 (1960): 15–29.

Al Sawda, Mahel. "The Rise and Transformation of Courtly Love: A Study in European Thought of Love." Diss. U of Essex, 1993.

Allen, Peter L. "Ars Amandi, Ars Legendi: Love Poetry and Literary Theory in Ovid, Andreas Capellanus, and Jean de Meun." *Exemplaria: A Journal of Theory in Medieval and Renaissance Studies* 1:1(1989): 181–205.

_____ . *The Art of Love: Amatory Fiction from Ovid to the Romance of the Rose.* Middle Ages Series. Philadelphia : U of Pennsylvania P, 1992.

Andersen-Wyman, Kathleen. "Andreas Capellanus on Love: Seduction, Subversion and Desire in a Twelfth-Century Text." Diss. U of California at Santa Cruz, 1993.

Andreas Capellanus. *The Art of Courtly Love [De amore].* Trans. J. J. Parry; ed. F. Locke. New York: F. Ungar, 1957.

Antonelli, Roberto. "The Birth of Criseyde—an Exemplary Triangle: 'Classical' Troilus and the Question of Love at the Anglo-Norman Court. *The European Tragedy of Troilus.* Ed. Piero Boitani. Oxford: Clarendon, 1989. 21–48.

Auerbach, Erich. "Camilla, or, The Rebirth of the Sublime." *Literary Language and Its Public in Late Latin Antiquity and in the Middle Ages.* Trans. R. Manheim. Bollingen Series LXXIV. New York: Pantheon Books, 1965.

_____ . "Figura." *Scenes from the Drama of European Literature.* Theory and History of Literature, vol. 9. Minneapolis: U of Minnesota Press.

Baldwin, John W. *The Language of Sex: Five Voices from Northern France around 1200.* The Chicago Series on Sexuality, History and Society. Chicago and London: U of Chicago Press, 1994.

_____ . *Aristocratic Life in Medieval France: The Romances of Jean Renart and Gerbert de Montreuil, 1190–1230.* Baltimore, MD : Johns Hopkins UP, 2000.

Baswell, Christopher. *Virgil in Medieval England: Figuring the Aeneid from the Twelfth Century to Chaucer.* Cambridge Studies in Medieval Literature, 24. Cambridge: Cambridge U P, 1995.

Bender, Karl-Heinz. "Beauté, mariage, amour: la genèse du premier roman courtois." *Amour, mariage et transgressions.* Ed. D. Buschinger and A. Crépin. Göppingen: Kümmerle Verlag, 1984. 173–183.

Benoit de Sainte-Maure. *Le roman de Troie.* Ed. L. E. Constans. SATF. Paris: Firmin-Didot, 1904–12.

Benton, John F. "Clio and Venus: An Historical View of Medieval Love." *The Meaning of Courtly Love.* Ed. F. X. Newman. Albany: State University of N. Y. Press, 1968. 19–42. This work also appears in: *Culture, Power and Personality in Medieval France.* Ed. T. Bisson. Ohio: Hambledon Press, 1991. 103–121.

Bernardus Silvestris. *The Commentary on the First Six Books of the Aeneid.* Ed. J. W. Jones and E. F. Jones. Lincoln, Nebraska; London: University of Nebraska Press, 1977.

Bezzola, Reto R. "Guillaume IX d'Aquitaine et les origines de l'amour courtois." *Romania* 66 (1940):145–237.

Birns, Nicholas. "The Trojan Myth: Postmodern Reverberations." *Exemplaria* 5 (1993): 45–78.

Blodgett, E. D. (ed.). *The Romance of Flamenca.* Garland Library of Medieval Literature. New York: Garland, 1995.

Boethius: The Consolation of Philosophy (with the English translation of "I. T." [1609]). 1918. Ed. and Revised H. F. Stewart. Loeb Classical Lib. Cambridge, Mass.: Harvard U Press, 1962.

Boulton, Maureen, and Barry McCann. "Lyric Insertions and the Reversal of Romance Conventions in Jean Renart's *Roman de la rose* or *Guillaume de Dole. Jean Renart and the Art of Romance: Essays on* Guillaume de Dole. Ed. Nancy Vine-Durling. Gainesville, FL : UP of Florida, 1997. 85–104.

Brackett, Maria Romagnoli. "Andreas Capellanus: Issues of Identity, Reception and Audience." Diss. Harvard U, 1996.

Bromwich, Rachel. "Celtic Dynastic Themes and the Breton Lays." *Etudes Celtiques* 9 (1961): 439–474.

_____ . *Trioedd Ynys Prydein: The Welsh Triads.* 1961. Rev. ed. Cardiff: U of Wales, 1978.

Bruckner, Matilda. "Romancing History and Rewriting the Game of Fiction: Jean Renart's *Rose* through the Looking Glass of *Partonopeu de Blois." The World and Its Rival: Essays on Literary Imagination in Honor of Per Nykrog.* Faux Titre: Etudes de Langue et Littérature Françaises. Ed. Kathryn Karczewska and Tom Conley. Amsterdam: Rodopi, 1999. 93–117.

Brundage, James A. *Law, Sex, and Christian Society in Medieval Europe.* Chicago; London: U of Chicago P, 1987.

Burgwinkle, William. "Knighting the Classical Hero: Homo/Hetero Affectivity in Eneas." *Exemplaria* 5:1 (1993): 1–43.

Chaucer, Geoffrey. *The Riverside Chaucer*. Ed. Larry D. Benson. 3rd ed. Boston: Houghton Mifflin, 1987.

Chenu, M.-D. *Nature, Man, and Society in the Twelfth Century*. Ed. and Trans. Jerome Taylor and Lester K. Little. Chicago: University of Chicago Press, 1968.

Cherchi, Paolo. *Andreas and the Ambiguity of Courtly Love*. Toronto Italian Studies. Toronto : U of Toronto P, 1994.

Chrétien de Troyes. *Cligès*. Ed. S. Gregory and C. Lutrell. Arthurian Studies 28. Cambridge, England: D. S. Brewer, 1993.

————. *Cligès. Œuvres complètes*. Ed. D. Poirion. Paris: Gallimard, 1994.

Colish, Marcia L. *Peter Lombard*. 2 vols. Leiden; New York; Köln: E. J. Brill, 1994.

Colliot, Régine. "Guillaume V d'Aquitaine et Le Comte Fouque dans *Girart de Roussillon*." *Bearn et Gascogne, de la réalité historique à la fiction romanesque*. 3e Colloque de Littérature Régionale. Pau: Université de Pau, 1983. 7–27.

Combarieu du Grès, Micheline de, and Gérard Gouiran, eds. and trans. *La Chanson de Girart de Roussillon*. Livre de Poche: Lettres Gothiques. [Paris?]: Librarie Générale Française, 1993.

————. "Le personnage d'Elissent dans *Girart de Roussillon*." *Studia Occitanica: in memoriam Paul Remy*. 2 vols. Ed. Hans-Erich Keller Kalamazoo: Med. Institute Pubs., 1986. II: 23–42.

Cooper, Linda. "L'Ironie iconographique de la coupe de Tristan dans *L'Escoufle*." *Romania* 104 (1983): 157–176.

Coppin, Joseph. *Amour et mariage dans la littérature française du Nord au Moyen-Age*. Bibliothèque Elzévirienne (n. s.). Paris: Librarie d'Argences, 1961.

Cormier, Raymond J. *One Heart One Mind: The Rebirth of Virgil's Hero in Medieval French Romance*. Romance Monographs, 3. University, Mississippi: Romance Monographs Inc., 1973.

Coudert, Allison. "Exemplary Biblical Couples and the Sacrament of Marriage." *Acta* 14 (1990): 59–83.

Crane, Susan. *Insular Romance: Politics, Faith and Culture in Anglo-Norman and Middle English Literature*. Berkeley: UC Press, 1986.

Curtius, Ernst Robert. *European Literature and the Latin Middle Ages*. 1948. Trans. W. R. Trask. Bollingen Series 36. Princeton, N. J.: Princeton U P, [1953] 1973. Trans. of *Europäische Literatur und lateinische Mittelalter*, 1948.

Darrigrand, Jean-Pierre (ed.). L'Amour courtois, des troubadours à Febus: *Flamenca*." *Actes des rencontres et communications Image/Imatge*. Orthez: Per Noste, 1995.

De Caluwé, Jacques. "L'amour et le mariage, moteurs seconds, dans la littérature épique française et occitane du XIIe siècle." *Love and Marriage in the Twelfth Century*. Ed. Willy Van Hoecke and Andries Welkenhuysen. Louvain (Belgium): Leuven U P, 1981. 171–182.

Dickey, Constance-L. "Deceit, Desire, Distance and Polysemy in Flamenca." *TENSO:-Bulletin of the Société Guilhem-IX* 11:1 (1995): 10–37.

Douglas, David C. *The Norman Fate 1100–1154*. Berkeley: UC Press, 1976.

Duby, Georges. *Love and Marriage in the Middle Ages*. Trans. Jane Dunnett. Chicago: U of Chicago P, 1994. Trans. *Mâle moyen âge*. Paris: Flammarion, 1988.

_____. *Medieval Marriage: Two Models from Twelfth-Century France*. Trans. E. Forster. Baltimore: John Hopkins U P, 1978.

_____. *The Knight, the Lady and the Priest: The Making of Marriage in Medieval France*. Trans. B. Bray. New York: Pantheon, 1983. Trans. of *Le chevalier, la femme et le prêtre*. Paris: Hachette, 1981.

Dumézil, Georges. *Mariages indo-européens*. Paris: Payot, 1979.

Ehrhart, Margaret J. *The Judgment of the Trojan Prince Paris in Medieval Literature*. Philadelphia: U of Pennsylvania Press, 1987.

Eley, Penny. "The Myth of Trojan Descent and Perceptions of National Identity: The Case of *Eneas* and the *Roman de Troie*." *Nottingham Medieval Studies* 35 (1991): 27–40.

Faral, Edmond. "Compte Rendu" [*Le roman de Troie*, Ed. L. Constans] *Romania* 42 (1913): 88–106.

_____. *Recherches sur les sources latines des contes et romans courtois du Moyen Age*. 1913. Paris: Honoré Champion, [1913] [1963] 1983.

Feimer, Joel N. "Jason and Medea in Benoît de Sainte-Maure's *Le Roman de Troie*: Classical Theme and Medieval Context." *Voices in Translation: The Authority of 'Oldes Bookes' in Medieval Literature*. Ed. D. M. Sinnreich-Levi and G. Sigal. New York: AMS Press, 1992. 35–51.

Ferroul, Yves. "La Passion selon Tristan et Iseut." *Et c'est la fin pour quoy sommes ensemble: Hommage à Jean Dufournet professeur à la Sorbonne Nouvelle: Littérature, histoire et langue du Moyen Age, I–III*. Nouvelle Bibliothèque du Moyen Age. Ed. Jean-Claude Aubailly and Emmanuele Baumgartner, *et. al.*. Paris: Champion, 1993. 571–78.

Foucault, Michel. *L'usage des plaisirs*. Histoire de la sexualité 2. Paris: Gallimard, 1984.

Fourrier, Anthime. *Le courant réaliste dans le roman courtois*. Paris: Nizet, 1960.

Frappier, Jean. "Structure et sens du *Tristan*: version commune, version courtoise." *Cahiers de civilisation médiévale* 6, no. 3 (1963): 255–280; 441–454.

_____. "Remarques sur la peinture de la vie des héros antiques dans la littérature française des 12e et 13e siècles." *L'humanisme médiéval dans les littératures romanes du XIIe au XIVe siècle*. Ed. Anthime Fourrier. Colloque de Strasbourg (1962). Paris: C. Klincksieck, 1964. 13–54.

Frazer, R. M. (trans.). *The Trojan War; The Chronicles of Dictys of Crete and Dares the Phrygian*. Indiana University Greek and Latin Classics. Bloomington: Indiana U P, 1966.

Gaudemet, Jean. *Le Mariage en Occident*. Les moeurs et le droit. Paris: Editions du Cerf, 1987.

Gaunt, Simon B. "Le pouvoir d'achat des femmes dans *Girart de Roussillon*." *Cahiers de Civilisation Médiévale* 33 (1990): 305–316.

_____. "Marginal Men, Marcabru and Orthodoxy: The Early Troubadours and Adultery." *Medium Aevum* 59 (1990): 55–71.

Godefroy, L. "Mariage" and "Le Mariage au temps des Pères." *Dictionnaire de Théologie Catholique*. Vol. 9:2. Ed. Vacant, Mangenot and Amann. Paris: Letouzey et Ané, 1927.

Gouiran, Gerard. "'Car tu es cavalliers e clers' (*Flamenca* v. 1899): Guilhem, ou le chevalier parfait." *Le Clerc au Moyen Age*. Aix-en-Provence: Centre Univ. d'Etudes et de Recherches Médiévales d'Aix, 1995. 197–214.

Graybill, Robert V. "Courts of Love: Challenge to Feudalism." *Essays in Medieval Studies* 5 (1988): 93–101.

Grimal, Pierre. *L'amour à Rome*. [Paris?]: Hachette, 1963.

Guynn, Noah D. "Historicizing Shame, Shaming History: Origination and Negativity in the *Eneas*." *Esprit Createur*, 39:4 (1999): 112–27.

Hackett, W. Mary. "L'élément courtois dans le vocabulaire de Girart de Roussillon." *La chanson de geste et le mythe carolingien. Mélanges René Louis*. 2 vols. Saint-Père-Sous-Vézelay: Musée Archéologique Régional, 1982. 2: 729–736.

Hanning, R. W. "The Social Significance of Twelfth-Century Chivalric Romance." *Medievalia et Humanistica* n. s. 3 (1972): 3–29.

Heintze, Michael. "Les techniques de la formation de cycles dans les chansons de geste." *Cyclification: The Development of Narrative Cycles in the Chansons de Geste and the Arthurian Romances*. Ed. B. Besamuca, *et. al.* Amsterdam; Oxford; New York: North-Holland, 1994.

Hexter, Ralph J. *Ovid and Medieval Schooling. Studies in Medieval School Commentaries on Ovid's* Ars Amatoria, Epistulae ex Ponto, *and* Epistulae Heroidum. Münchener Beiträge zur Mediävistik und Renaissance-Forschung 38. München: Arbeo-Gesellschaft, 1986.

Hugh of Saint Victor. *Hugh of Saint Victor on the Sacraments of the Christian Faith (De Sacramentis)*. Trans. Roy J. Deferrari. Cambridge, MA: The Medieval Academy of America, 1951.

Hughes, Diane Owen. "From Brideprice to Dowry." *Journal of Family History* 3 (1978): 262–296.

Hunt, Tony. "Abelardian Ethics and Béroul's *Tristan*." *Romania* 98 (1977): 501–540.

Jeanjean, Henri. "Flamenca: A Wake for a Dying Civilization?" *Parergon* 16:1 (1998): 19–30.

Jewers, Caroline A. "Sentimental Education: The *Roman de Flamenca* and the Renaissance of the Ovidian Hero." *Proceedings of the Medieval Association of the Midwest, II*. Ed. Mel Storm. Emporia, KS : Emporia State U, 1993. 58–70.

_____ . "Fabric and Fabrication: Lyric and Narrative in Jean Renart's Roman de la Rose." *Speculum* 71:4 (1996): 907–24.

Jodogne, Omer. "Le caractère des oeuvres *antiques* dans la littérature française du 12e et du 13e siècle." *L'humanisme médiéval dans les littératures romanes du XIIe au XIVe siècle*. Ed. Anthime Fourrier. Colloque de Strasbourg (1962). Paris: C. Klincksieck, 1964. 55–85.

Jones, Rosemary. *The Theme of Love in the Romans d'Antiquité*. Dissertation Series 5. London: The Modern Humanities Research Association. 1972.

Jonin, Pierre. *Les personnages féminins dans les romans français de Tristan au XIIe siècle.* Gap: Ophrys, 1958.

Karnein, Alfred. "De Amore in volkssprachlicher Literatur: Untersuchungen zur Andreas-Capellanus-Rezeption in Mittelalter und Renaissance." *Germanisch-Romanische-Monatsschrift.* 4: supp. (1985): 1–337.

Kay, Sarah. "Courts, Clerks, and Courtly Love." *The Cambridge Companion to Medieval Romance.* Ed. Roberta L. Krueger. Cambridge, England: Cambridge U P, 2000. 81–96.

_____ . "The Contradictions of Courtly Love and the Origins of Courtly Poetry: The Evidence of the Lauzengiers." *Journal of Medieval and Renaissance Studies.* 26:2 (1996): 209–253.

_____ . "The Tristan Story as Chivalric Romance, Feudal Epic and Fabliau." *The Spirit of the Court: Selected Proceedings . . .International courtly literature society.* Ed. Glyn S. Burgess and R. A. Taylor. Dover, N. H.: D. S. Brewer, 1985. 185–95.

Kenaan, Vered-Lev. "The Erotodidactic Discourse of Ovid and Andreas Capellanus." Diss. Yale U, 1995.

Köhler, Erich. "Observations historiques and sociologiques sur la poésie des troubadours." *Cahiers de civilisation médiévale* 7 (1964): 27–51.

_____ . *L'aventure chevaleresque: Idéal et réalité dans le roman courtois.* Paris: Gallimard, 1974. Trans. of *Ideal und Wirklichkeit in der Höfischen Epik,* 1956.

Labbé, Alain. "La Comtesse Berthe dans Girart de Roussillon: 'L'Amour et la vie d'une femme.'" *Charlemagne in the North. Proc. of the 12th Internat. Conf. of the Soc. Roncesvals* (1991). Ed. Philip E. Bennett; Anne Elizabeth Cobby, and Graham A. Runnalls. Edinburgh, 1993. 319–33.

_____ . "L'Espace littéraire et politique de Girart de Roussillon: Une Géographie heritée de l'histoire et investie par la poésie." *Provinces, regions et terroirs au moyen age. Actes du colloque de Strasbourg,* Nancy : PU de Nancy, 1991. 313–33.

Lafont, Robert. "*Girart de Roussillon*: Un Texte occitan." *De l'aventure épique à l'aventure romanesque. Mélanges offerts à Andre de Mandach par ses amis, collègues et élèves.* Ed. Jacques Chocheyras. Bern, Switzerland: Peter Lang, 1997. 29–50.

Laurie, H. C. R. "Eneas and the Doctrine of Courtly Love." *Modern Language Review* 54 (1969) 283–294.

Lazar, Moshé. *Amour courtois et* fin'amors *dans la littérature du XIIe siècle.* Paris: C. Klincksieck, 1964.

Le Bras, G. "La Doctrine du mariage chez les Théologiens et les Canonistes depuis l'an mil." *Dictionnaire de Théologie Catholique.* Vol. 9:2. Ed. Vacant, Mangenot and Amann. Paris: Letouzey et Ané, 1927.

_____ , Ch. Lefebvre, and J. Rambaud. *Histoire du droit et des institutions de l'Eglise en Occident.* Vol. 7: *L'Age classique* (1140–1378): *Sources et théorie du droit.* Paris: Sirey, 1965.

Le Gentil, Pierre. "*Girart de Roussillon,* sens et structure du poème." *Romania* 78 (1957): 328–389 and 463–510.

_____ . "La légende de Tristan vue par Béroul et Thomas: Essai d'interprétation." *Romance Philology* 7 (1953–4): 111–129.

Le Goff, Jacques. *Pour un autre Moyen Age: Temps, travail et culture en Occident.* Paris: Gallimard, 1977.

Lewes, Ulle Erika. "*The Life in the Forest: The Influence of the St. Giles Legend on the Courtly Tristan Story.*" Chattanooga, Tennessee: Tristania Monographs, 1978.

Loomis, Roger Sherman. *Arthurian Legends in Medieval Art.* (Monograph Series, IX) New York: Modern Language Association of America, 1938.

Louis, René. *De l'histoire à la légende.* 2 vols. Auxerre: Imprimerie Moderne, 1946–47.

Loyd, James; Leon-de-Vivero, Virginia. "The Artful Rejection of Love from Ovid to Andreas to Chartier." *Modern Language Studies* 9:2 (1979): 46–52.

Luce-Dudemaine, Marie Dominique. "Un Nouvel Art d'aimer, la contestation des valeurs courtoises dans Flamenca." *Revue des Langues Romanes* 92:1 (1988): 61–75.

Lulua, Abdul Wahid. "Courtly Love: Arabian or European?" *Proceedings of the XIIth Congress of the International Comparative Literature Association/Actes du XIIe congres de l'Association Internationale de Litterature Comparee: Munchen 1988 Munich, IV: Space and Boundaries of Literature/Espace et frontieres de la litterature.* Ed. Roger Bauer and Douwe Fokkema. Munich: Iudicium, 1990. 390–396.

Lumiansky, R. M. "Structural Unity in Benoît's *Roman de Troie.*" *Romania* 79 (1958): 410–424.

Mackey, Louis. "Eros into Logic: The Rhetoric of Courtly Love." *The Philosophy of (Erotic) Love.* Ed. Robert C. Solomon; Kathleen M. Higgins; Lawrence : UP of Kansas, 1991: 336–51.

Maddox, Donald L. "Pseudo-Historical Discourse in Fiction: *Cligés.*" *Essays in Early French Literature Presented to Barbara M. Craig.* Ed. N. J. Lacy and J. C. Nash. York, South Carolina: French Lit. Pubs., 1982: 9–24.

_____ . "Critical Trends and Recent Works on the *Cligès* of Chrétien de Troyes." *Neuphilologische Mitteilungen* 74 (1973): 730–745.

Mandach, Andre de, and, Eve Marie Roth. "Le Triangle Marc-Iseut-Tristan: Un Drame de double inceste." *Etudes-Celtiques* 23 (1986): 193–213.

Mandach, Andre de, and, Eve Marie Roth. "Le Triangle Marc-Iseut-Tristan: Un Drame de double inceste." *Etudes-Celtiques* 23 (1986): 193–213.

Markale, Jean. *Women of the Celts.* 1972. Trans. A Mygind, C. Hauch, and P. Henry. Rochester, VT: Inner Traditions Int., 1986.

Matoré, Georges. *Le Vocabulaire et la société médiévale.* Paris: P U F, 1985.

McCone, Kim. *Pagan Past and Christian Present in Early Irish Literature.* Maynooth Monographs 3. Naas, Co. Kildare, Ireland: An Sagart 1990. Rpt. 1991.

Monson, Donald A. "Andreas Capellanus and the Problem of Irony." *Speculum* 63:3 (1988): 539–572.

————. "Andreas Capellanus's Scholastic Definition of Love." *Viator* 25 (1994): 197–214.

————. "Auctoritas and Intertextuality in Andreas Capellanus' *De Amore.*" *Poetics of Love in the Middle Ages: Texts and Contexts.* Ed. Moshé Lazar and Norris J. Lacy. Fairfax, VA : George Mason U P, 1989. 69–79.

Nelson, Deborah. "The Public and Private Images of *Cligès*' Fénice." *Reading Medieval Studies* 7 (1981): 81–88.

Newman, F. X., ed. *The Meaning of Courtly Love.* Albany: State University of N. Y. Press, 1968.

Nichols, Stephen G. "Rewriting Marriage in the Middle Ages." *Romanic Review* 79:1 (1988): 42–60.

Nickolaus, Keith A. "Social Class Ideology and Medieval Love: Marriage Fictions in Girart de Roussillon." *Romance Languages Annual* 8 (1996): 84–90.

Noble, Peter. *Love and Marriage in Chrétien de Troyes.* Cardiff: U of Whales P, 1982.

Nolan, Barbara. "Ovid's *Heroides* Contextualized: Foolish Love and Legitimate Marriage in the *Roman d'Eneas.*" *Mediaevlia* 13 (1987): 157–187.

Noonan, John T., Jr. "Marital Affection in the Canonists." *Studia Gratiana* (XII) 1967: 479–510.

————. "Power to Choose." *Marriage in the Middle Ages. Viator* 4 (1973): 419–434.

O'Callaghan, Tamara. "Faith and Love Imagery in Benoit de Sainte-Maure's *Roman de Troie,* John Gower's *Confessio Amantis,* and Geoffrey Chaucer's *Troilus and Criseyde.*" Diss. U of Toronto, 1995.

Ovid. *Heroides.* Trans. Grant Showerman. Loeb Classical Library. Cambridge: Harvard U P, 1921.

————. *Metamorphoses.* 1916. Trans. Frank Justus Miller. Loeb Classical Library. Cambridge: Harvard U P, 1984.

————. *The Art of Love and Other Poems.* Trans. J. H. Mozley. Loeb Classical Library, 232. Cambridge, MA: Harvard U P, 1939.

Paden, William D., Jr. "The Troubadour's Lady: Her Marital Status and Social Rank." *Studies in Philology* 72 (1975): 28–50.

Painter, Sidney. "The Family and the Feudal System in Twelfth-Century England." *Feudalism and Liberty: Articles and Addresses of SidneyPainter.* Ed. F. A. Cazel, Jr. Baltimore: Johns Hopkins Press, 1961. 195–219.

Partner, Nancy. *Serious Entertainments: The Writing of History in Twelfth-Century England.* Chicago: U of Chicago Press, 1977.

Partonopeus de Blois. Anciens monuments de l'histoire et de la langue françaises IV. Paris: Crapelet, 1834. Geneva: Slatkine Reprints, 1976.

Patch, Howard R. *The Goddess Fortuna in Mediaeval Literature.* Cambridge: Harvard U P, 1927.

Patterson, Lee. "Virgil and the Historical Consciousness of the Twelfth Century: The *Roman d'Eneas* and *Erec et Enide.*" *Negotiating the Past: The Historical*

Understanding of Medieval Literature. Madison: University of Wisconsin Press, 1987. 157–195.

Payen, Jean Charles. "La 'Mise en Roman' du mariage dans la littérature française des XIIe et XIIIe siècles: de l'évolution idéologique à la typologie des genres." *Love and Marriage in the Twelfth Century.* Ed. Willy Van Hoecke and Andries Welkenhuysen. Louvain (Belgium): Leuven U P, 1981. 219–235.

Perugi, Maurizio. *Saggi di linguistica trovadorica: Saggi su Girart de Roussillon, Marcabruno, Bernart de Ventadorn, Raimbaut d'Aurenga, Arnaut Daniel e sull'uso letterario de oc e oil nel Trecento italiano.* Romanica-et-Comparatistica. Tubingen : Stauffenburg, 1995.

Petit, Aimé. "Enéas dans le Roman d'Enéas." *Le Moyen-Age: Révue d'Histoire et de Philologie* 96:1 (1990): 67–79.

Pinder, Janice M. "The Intextuality of Old French Saints' Lives: St. Giles, St. Evroul and the Marriage of St. Alexis." *Parergon* 6A (1988): 11–21.

Poirion, Daniel. "De l'*Enéide* à l'*Enéas*: mythologie et moralisation." *Cahiers de Civilisation Médiévale* 19 (1976), 213–229.

_____ . "L'Ecriture épique: du sublime au symbole." *Relire le* Roman d'Enéas. Ed. Jean Dufournet. Collection Unichamp 8. Paris: Honoré Champion, 1985. i–xiii.

Press, A. R. "The Adulterous Nature of *fin'amors*: A Re-examination of the Theory." *Forum for Modern Language Studies* 6 (1970): 327–341.

Quinn, John Francis, C.S.B. "Saint Bonaventure and the Sacrament of Marriage." *Franciscan Studies* 34 (1974): 101–43.

Robertson, D. W. "The Concept of Courtly Love as an Impediment to the Understanding of Medieval Texts." *The Meaning of Courtly Love.* Ed. F. X. Newman. Albany: State U of New York P, 1968. 1–18.

Rollo, David Iain. "*Flamenca*: Intergeneric Copula as Political Allegory." Dissertation Abstracts International 49:7 (1989): 1798A.

Salverda de Grave, J. J., ed. *Eneas, roman du XIIe siècle.* CFMA, 44, 62. Paris: H. Champion, 1925–29.

Schnell, Rudiger. *Andreas Capellanus: Zur Rezeption des romischen und kanonischen Rechts in* De Amore. Munstersche-Mittelalter-Schriften 46. Munich : Fink, 1982.

Sheehan, Michael M. "Choice of Marriage Partner in the Middle Ages: Development and Mode of Application of a Theory of Marriage." *Studies in Medieval and Renaissance History*, n. s. 1 (1978): 3–33.

Shirt, David J. "*Cligès*: A Twelfth-Century Matrimonial Case-Book?" *Forum for Modern Language Studies* 18 (1982): 75–89.

Simons, Penny and Penny Ely. "The Prologue to *Partonopeus de Blois*: Text, Context and Subtext," *French Studies*, 49 (1995): 1–16.

Spence, Sarah. *Rhetoric of Reason and Desire: Vergil, Augustine and the Troubadours.* Ithaca and London: Cornell U Press, 1988.

Spiegel, Gabrielle M. "Genealogy: Form and Function in Medieval Historical Narrative." *History and Theory* 22 (1983): 43–53.

Subrenat, Jean. "Le climat social, moral, religieux du *Tristan* de Béroul." *Le Moyen Age*, 82 (1976): 219–261.

Suhamy, Henri. "Le Mariage imaginaire, ou l'humanisme en question: Quelques hypothèses sur Troilus and Cressida." *Etudes Anglaises: Grande Bretagne, Etats-Unis* 44:1 (1991): 1–14.

Teperman, Andrée Graciela. "Genres littéraires et contexte culturel: l'évolution des structures du mariage dans l'épopée et le roman de la France médiévale." Diss. UC Berkeley, 1986.

Thaner, F. (ed.). *Summa Magistri Rolandi* [Pope Alexander III]. 1874. Innsbruck: Scientia Verlag Aalen, 1962.

Tobler, A., E. Lommatzsch. *Altfranzösisches Wörterbuch*. Vol. 2. 1956.

Tristan et Iseut: Les poèmes français; La saga norroise. Ed. and trans. D. Lacroix and P. Walter. Livre de Poche: Lettres Gothiques. [Paris?]: Librarie Générale Française, 1989.

Viau, Martine; Rochedereux, Evelyne. "Aliénor d'Aquitaine et la littérature courtoise." *Lesbia Magazine* 156 (1997): 24–30.

Vine-Durling, Nancy (ed.). *Jean Renart and the Art of Romance: Essays on Guillaume de Dole*. Gainesville, FL : U P of Florida, 1997.

_____ . "The Seal and the Rose: Erotic Exchanges in *Guillaume de Dole*." *Neophilologus*, 77:1 (1993): 31–40.

Virgil. *Aeneid*. 1916. Trans. H. Rushton Fairclough. 2 vols. Loeb Classical Library. Cambridge, MA: Harvard U P 1978.

Warren, Wilfred Lewis. *Henry II*. Berkeley and Los Angeles: U of California P, 1973.

Williams, Andrew. "Clerics and Courtly Love in Andreas Capellanus' *The Art of Courtly Love* and Chaucer's *Canterbury Tales*." *Revista Alicantina de Estudios Ingleses* 3 (1990): 127–36.

Williams, Andrew. "Clerics and Courtly Love in Andreas Capellanus' *The Art of Courtly Love* and Chaucer's *Canterbury Tales*. *Revista-Alicantina-de-Estudios-Ingleses* 3 (1990): 127–36.

Yunck, John A. (Trans. and Ed.) *Eneas, A Twelfth-Century French Romance*. Records of Civilization: Sources and Studies 93. New York: Columbia University Press, 1974.

Zak, Nancy C. "Modes of Love in *Flamenca*: Legitimate/Illegitimate, Vital/Sterile/Human/Inhuman." *Middle Ages: Texts and Contexts*. Ed. Moshé Lazar and Norris J. Lacy(ed.). Fairfax, VA : George Mason U P, 1989. 43–51.

Index